ORDER NUMBER EA-DCC

DC CIRCUITS

By Stan Rosen

International Standard Book Number 0-89100-121-2
For sale by: IAP, Inc., A Hawks Industries Company
Mail To: P.O. Box 10000, Casper, WY 82602-1000
Ship To: 7383 6WN Road, Casper, WY 82604-1835
(800) 443-9250 ❖ (307) 266-3838 ❖ FAX: (307) 472-5106

HBC0293 Printed in the USA

Printed in the United States of America

Library of Congress Cataloging-in-Publication Data

Rosen, Stan 1939-
D.C. Circuits

1. Electric Circuits — Direct Current.
I. Aviation Maintenance Publishers.
II. Title.
TK454.15.D57R67 621.319'12 80-12328

ISBN 0-89100-121-2

To:

My wife, Nancy, and my children,
Sacha and Mimi,
in appreciation for their
guidance, aid, and inspiration.

Table of Contents

Preface

This book on *DC Circuits* for avionics technicians, is one of a series of specialized texts prepared for the study of aviation electronics (avionics).

It is the purpose of this series to provide an integrated, classroom-oriented program stressing the practical application of theory to the installation and maintenance of avionics equipment. In the earlier works of this series, the establishment of a strong background and flexible approach are stressed, permitting immediate entrance at either a first-level exposure for students with a limited background, or a second-level exposure for more experienced personnel. These introductory works provide training in DC circuits, AC circuits, semiconductor theory and application, digital logic, microwaves, test equipment, and maintenance practices.

With the knowledge obtained from the introductory portion of the series, the student is able to proceed to the intermediate level. This level consists of a series of books and audio-visual presentations which cover the circuits which may be considered the building blocks of all avionics equipment, such as power supply circuits, transistor amplifier and switching circuits, oscillator and timing circuits, etc. Since these circuits or variations of them are the base of all more complex systems, a component function and failure analysis approach is taken in this part of the series. Exercises and examples from functional equipment enable the student to gain experience not only with the detailed theory of operation of these common circuits, but with the symptoms of particular component failure as well.

The third portion of the series consists of books giving in-depth treatment of each of the systems, on board or associated with, aircraft. Each book traces the history of the development of a particular system and includes both a generalized analysis of presently operating systems, as well as a detailed stage-by-stage analysis of the operation and troubleshooting techniques used in a popular or typical unit.

Acknowledgements

A word of thanks is expressed to the Wilcox Electric Co., Kansas City, Missouri, for permission to use a number of illustrations from their electronic units. It is to be understood that these are used for illustrative purposes only, and all rights, including patent and reproduction, remain with their original holders.

If you have any questions or comments regarding this manual, or any of the many other textbooks offered by IAP, simply contact: Sales Department, IAP, Inc.; Mailing Address: P.O. Box 10000, Casper, WY 82602-1000; Shipping Address: 7383 6WN Road, Casper, WY 82604-1835; or call toll free: (800) 443-9250; International, call: (307) 266-3838.

Introduction

This book is designed to provide basic theory and practical information for the avionics technology student. Because it combines both theory and practical information, this book is useful to the beginner, whose experience in electronics is limited, or the experienced technician with training in another field of electronics. Naturally, the way the book is used will depend on the individual's level of experience.

DC Circuits for Avionics Technicians can be used without refresher material by someone who has taken high school algebra, trigonometry, and physics, and is familiar with the basic fundamentals of electricity. Careful attention to the self-test and study questions at the end of each section is needed. These questions are designed to help the individual recall and understand the material in each section. If one is unable to answer all of the questions, the material should be reviewed.

SECTION I

Electricity and the Atom

A. Introduction

We all know that if a glass of water is standing in direct sunlight, the water will be heated. What happens is that the energy contained in the sunlight performs work upon tiny particles of water called molecules, causing them to rush about in a more agitated manner. It is this increase in the activity of the water molecules that we perceive as an increase in temperature. When we put the water in the sunlight, we are using energy to do work.

Although we do not often put a glass of water in the sun to heat it, as members of a highly technological society, we are constantly using and controlling energy to do work. For example, the solar energy that was stored in growing plants millions of years ago, and then converted to petroleum by great pressure deep in the earth, is used to lift a modern aircraft. The aircraft engine, therefore, may be seen as a tool for tapping the solar and geological energy stored in petroleum, controlling it, and applying it to do the work of powering an aircraft.

In a similar fashion, the electronic equipment and circuits we shall examine in this and other volumes in this series, are devices for *tapping*, *controlling*, and *applying* electrical energy, or electricity, to the work required by the complex nature of flight. This fairly simple principle, as you will see, is the base for our course of study. If we keep this idea firmly in mind, many complex devices and difficult-to-understand circuits can be mastered. The key questions to ask when beginning to study a particular device or circuit are:

1) How does this device or circuit make use of or control electrical energy?

2) How does this device or circuit contribute to the work of flying an aircraft?

Although not much older than some of those presently at work in it, the field of aviation electronics, or *avionics* as it is usually called, is one of the most demanding specialties in terms of the level of theoretical and practical knowledge required of the technician. It is the job of the avionics technician to install and maintain the electronic devices upon which the lives of countless passengers and crew members depend. The fact that human lives are at risk, coupled with the fact that avionics is generally among the first applications of a new concept or device, makes it necessary for the avionics technician to understand the nature of the energy with which he is dealing, as well as the various methods used to make it perform a particular function.

B. The Atom and Its Parts

The Greek philosopher Democritus of Abdera originated the word *atom* in about 430 B.C. He presented the idea that indivisible atoms were the basis of all material things, but this idea remained little more than an abstract philosophical concept until the seventeenth century.

Beginning at that time, research into the nature of gases under pressure, magnetism, and, somewhat later, static electricity, gradually produced a body of observations that could be explained only by reference to atoms and to parts of atoms.

Among the earliest of these researches were those of the Englishman Robert Boyle, whose book, *The Sceptical Chemist* (published in 1661), gave the results of some of his experiments with gases and presented the notion of a chemical *element*. According to Boyle, a chemical element was one of the simple substances of which the world was formed. It could not be broken down into simpler chemical entities.

A hundred and fifty years later Boyle's notions led to Dalton's more modern atomic theory.

Building on the careful work in chemical analysis of the seventeenth and eighteenth centuries, Dalton was able to show that each chemical element had to be composed of atoms of the same weight and that different elements, copper and silver, for example, differed because they were composed of atoms of different weight.

At about the same time that chemistry was advancing toward an atomic theory, the scientific study of electricity began. From the early work of the German physicist Otto von Guericke, whose experiments with air pumps had inspired the researches of Robert Boyle, to Benjamin Franklin and his famous kite, slow progress was made.

Franklin thought there was an electric "fluid" which could be added to or removed from substances by rubbing or by other means. The adding or removing of electric fluid resulted in an electric *charge* upon the object from which the fluid was removed or to which it was added.

Franklin called those objects to which he assumed the fluid had been added, *positively* charged. Those objects from which the fluid had supposedly been removed, such as a rubber rod rubbed with woolen cloth, he called *negatively* charged. Objects with the same charge, it was noted, repelled each other. But a force of attraction, diminishing rapidly as the distance between them increased, tended to draw objects with unlike charges together. Somehow, the space around a charged body was changed. This could be detected by bringing a second charged body into the neighborhood. Thus arose the concept of an *electric field.*

Franklin thought that when a wire is connected between oppositely charged bodies, the electric fluid would flow from the positively to the negatively charged body through the wire. This grew into the concept of a *current* or flow from positive to negative.

Today we know that Franklin's theories were somewhat in error. When oppositely charged bodies are connected by a wire, the flow of electrons is from the negatively charged body to the positively charged one, for a negative charge is formed by *adding* electrons to a body, while a positive charge is formed by removing electrons from it.

Unfortunately, this realization came well after many important discoveries in electricity were made, and Franklin's notion of current flow from positive to negative became a permanent fixture of electrical theory, in spite of the fact that most electronic circuits are based upon the movement of electrons.

A number of textbooks attempted to correct this error by speaking of current flow from negative to positive, the direction in which the electrons flow. For the most part, these attempts succeeded only in introducing further confusion. This book, as well as the others in this series, will distinguish between electron flow, which takes place from negatively charged bodies toward positively charged ones, and conventional current flow, which can be viewed as taking place in the opposite direction. When we come to discuss the behavior of semiconductors, it will be seen to be convenient to be able to speak of two opposite flows.

Before the experiments performed by the British physicists Thomson and Rutherford around the beginning of the twentieth century, it was generally thought that the atom was a solid, indestructible body. Thomson's and Rutherford's experiments led to the very different notion that the atom was composed of a relatively massive *nucleus* surrounded by a swarm of lighter particles spinning around it (Fig. 1-1).

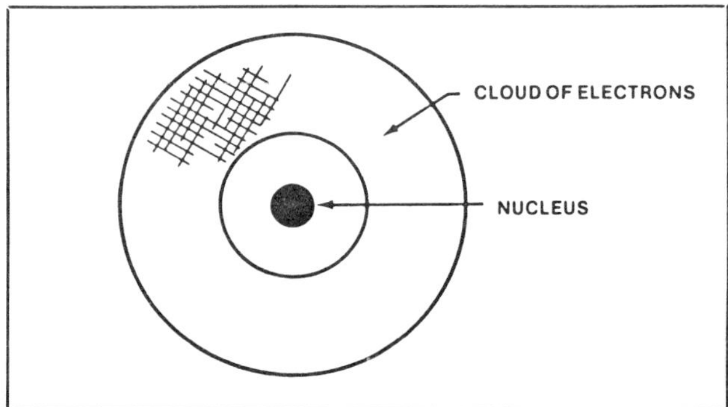

Fig. 1-1 The Rutherford model of the atom (1911) showed it as a massive nucleus surrounded by a swarm of lighter particles.

Although not completely in agreement with all that was known at the time, Rutherford's conception of the atom made it possible to explain the body of observations about the world which had been loosely grouped under the headings *electrical, chemical,* or *magnetic.*

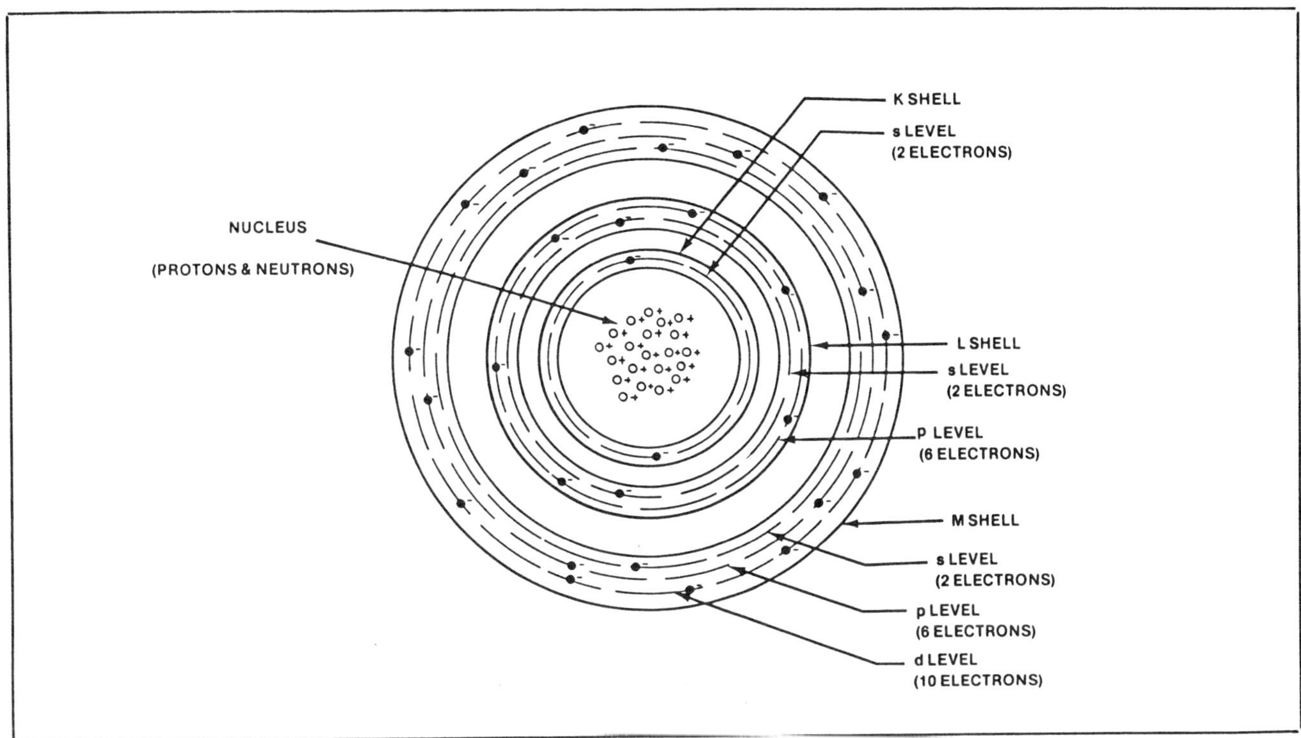

Fig. 1-2 The Bohr model of the atom, proposed in 1913, still serves for most purposes.

Rutherford noted that the atom was held together by the electric field that existed between the nucleus, which had a positive charge, and the surrounding electrons, which possessed negative charges. Electrical charge, the phenomenon that had been studied under many different forms, could be linked to the atom.

Twenty-five years ago, before the tremendous rise of solid-state electronics, an introductory textbook could have stopped after presenting Rutherford's model of the atom and defining the electron. An understanding of solid-state electronics requires two further steps, the additions to the Rutherford model of the atom suggested by Niels Bohr, and an investigation of the way in which charge carriers operate when atoms are organized in substances.

C. Bohr's Model of the Atom

In 1913, the Danish physicist Niels Bohr suggested a number of additions to Rutherford's atomic theory. Although some sixty-five years have passed since Bohr made his views public, his model of the atom has not been improved upon insofar as the needs of the avionics technician are concerned. Taken together, Rutherford's and Bohr's theories call for an atomic nucleus made up of one or more positively charged bodies called *protons*. Revolving around this nucleus at specific distances are negatively charged *electrons*, equal in number to the protons which form the nucleus.

The electrons are much lighter than protons, approximately 1850 times less massive. The negative charge of the electrons in each atom balances the equal but opposite charge of the protons; the atom as a whole has no net electrical charge. Since bodies with the same charge repel each other, the repelling force would tend to break the nucleus apart. However, this is compensated for by the presence of a number of uncharged bodies in the nucleus. These bodies, called *neutrons*, about equal in mass to the protons, seem to serve as a glue to hold the nucleus together.

There must be a minimum number of neutrons present for a given number of protons or the nucleus will be unstable and break apart, releasing a relatively vast amount of energy for such a small object. Only the nucleus of the ordinary hydrogen atom, composed of one lone proton, does not need any neutron "glue."

The basic differences between the Rutherford and Rutherford-Bohr models of the atom which concern the avionics technician are pictured in Fig.

1-2. As may be seen in the figure, Bohr suggested that the path of any particular electron around its nucleus is strictly determined by the number of electrons present and the energy level of each electron.

The paths are organized into *shells*. Each shell contains a number of distinct paths which are now called *energy levels*. Each energy level may contain only up to a maximum number of electrons.

As seen in Fig. 1-2, the shells, beginning with the one closest to the nucleus, are labelled with upper-case letters K, L, M, N, etc. The K shell has only one energy level, s, and can contain only two electrons. The simplest atom, that of hydrogen, whose nucleus contains but one proton, will have a single electron in the s energy level of the K shell. (This is true, as we shall see below, only when no additional energy has been added to the electron, but for the moment it can be accepted as a simplification.)

The next most complex atom is the helium atom. Its nucleus is made up of two protons and two neutrons, with two electrons in the s level of the K shell. This is the maximum capacity of the s level.

The next element, lithium, has three protons in its nucleus. The first two of its three electrons fill the s level of the K shell. Since the K shell has only one energy level, the third lithium electron occupies the s level of the second or L shell. The L shell possesses two energy levels, the s level with a capacity of two electrons, and the p level with a capacity of 6, so that the L shell will contain 8 electrons before the next shell is started.

The element whose number of electrons fills the L shell is neon, with ten electrons in all — two in the s level of the K shell, two in the s level of the L shell, and six in the p level of the L shell. It is an interesting and important point that the neon atom which just fills its L shell, like the helium atom which just fills its K shell, shows very little chemical activity and clings to its electrons strongly. The filled shell seems to be a preferred, stable state.

Lithium, on the other hand, which has filled its K shell and has a single electron in the L shell enters eagerly into chemical activity with other elements. The outer electron, called the *valence electron*, can be given up to another atom which needs a single electron to complete a shell.

For example, lithium forms a compound with hydrogen, called lithium hydride. In this compound, the outer lithium electron is transferred to the K shell of the hydrogen atom, making the charge of the lithium atom as a whole +1 and the charge of the hydrogen atom −1.

These charges hold the atoms together to form a single molecule of the compound lithium hydride. Compounds held together in this fashion are said to have an *electrovalent* or *ionic bond* (Fig. 1-3).

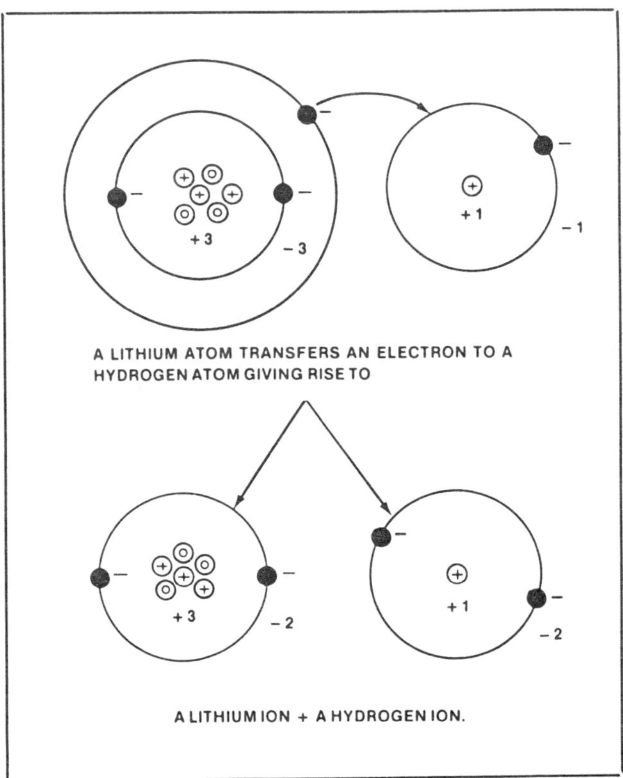

Fig. 1-3 Electrovalent bond or ionic bond — a compound formed by the electrical attraction between the positive lithium ion and the negative hydrogen ion.

In nature, these substances, such as ordinary table salt, often form crystal structures consisting of intermingled charged atoms or *ions*, as they are called.

Another sort of bonding can take place when an outermost shell is partially filled, as in the case of the elements carbon, silicon, and germanium, elements which are quite important to the electronics industry. An atom of carbon, for example, will fill its outermost shell by *sharing* electrons with other atoms.

In the familiar diamond, each carbon atom shares an electron pair with each of three neighboring carbon atoms to form what is called a *covalent bond* (Fig. 1-4).

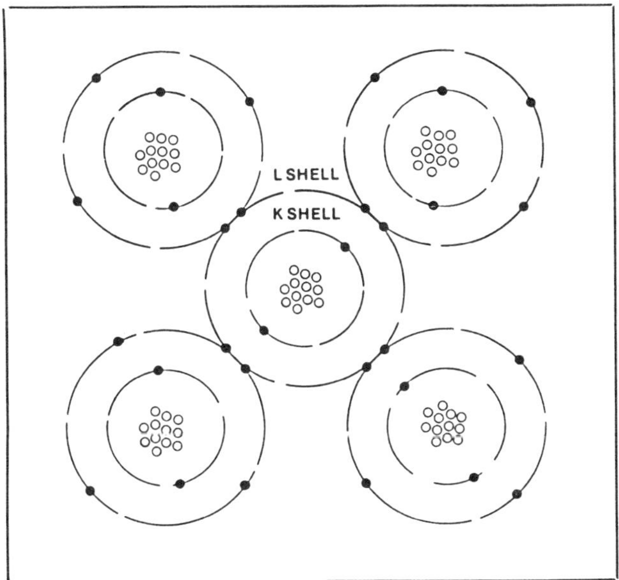

Fig. 1-4 Covalent bond — the bonding of carbon atoms which share electrons.

Since covalent bonding brings the outermost shells of the atoms into contact, the bond is usually closer and stronger than an electrovalent bond.

Table 1-1 shows the energy levels present in each of the shells and the maximum number of electrons allowed in each level. As we have seen in the case of lithium, the number of valence electrons, those in the outermost energy level or levels, determines most of the physical and chemical properties of an element.

This is summarized in the familiar periodic table of the elements arranged according to the number of protons in the nucleus (the atomic number) and into vertical columns according to the number of valence electrons (up to eight). Thus, we may expect elements in the same vertical column to be chemically similar.

In the periodic table (Fig. 1-5), each box represents a single element and gives the name of the element, its chemical abbreviation (normally the first two letters of its Latin name), its atomic number, which is the number of protons in its nucleus, and the arrangement of its electrons in shells and energy levels.

Table 1-1

Shells and Energy Levels

Shell	Energy Levels	Maximum Number of Electrons
K	s	2
L	s p	2 6
M	s p d	2 6 10
N	s p d f	2 6 10 14
O	s p d f g	2 6 10 14 18
P	s p d f g h	2 6 10 14 18 22
Q*	s	2

*The various higher energy levels of the Q shell are not shown, since there is no atom with enough electrons to fill them.

As you can see, in the elements after argon, atomic number 18, the distribution of electrons becomes complicated. Instead of filling the d level of the M shell, the nineteenth electron of potassium, element number 19, occupies the s level of the N shell. With the element number 21, scandium, the electrons begin filling in the d level of the M shell, giving rise to a series of *transition elements* with similar chemical and physical properties. Note that these transition elements have the same number of valence electrons.

This first group of transition elements includes the majority of those elements whose properties make them useful in electric or magnetic circuits. Nearly all the elements which are commonly used to conduct electricity or for their magnetic properties fall into one of the transition families.

It should be noted that the two elements used in transistors — silicon, number 14, and germanium, number 32 — both fall into the column of those elements possessing four valence electrons.

By now you may have been wondering why the electron paths in the various shells are called energy levels. The reason for this lies at the heart of Bohr's conception of the atom. Bohr suggested that the particular path that an electron takes depends not only upon the requirements of the electron shells, but also upon the amount of

PERIODIC TABLE OF ELEMENTS

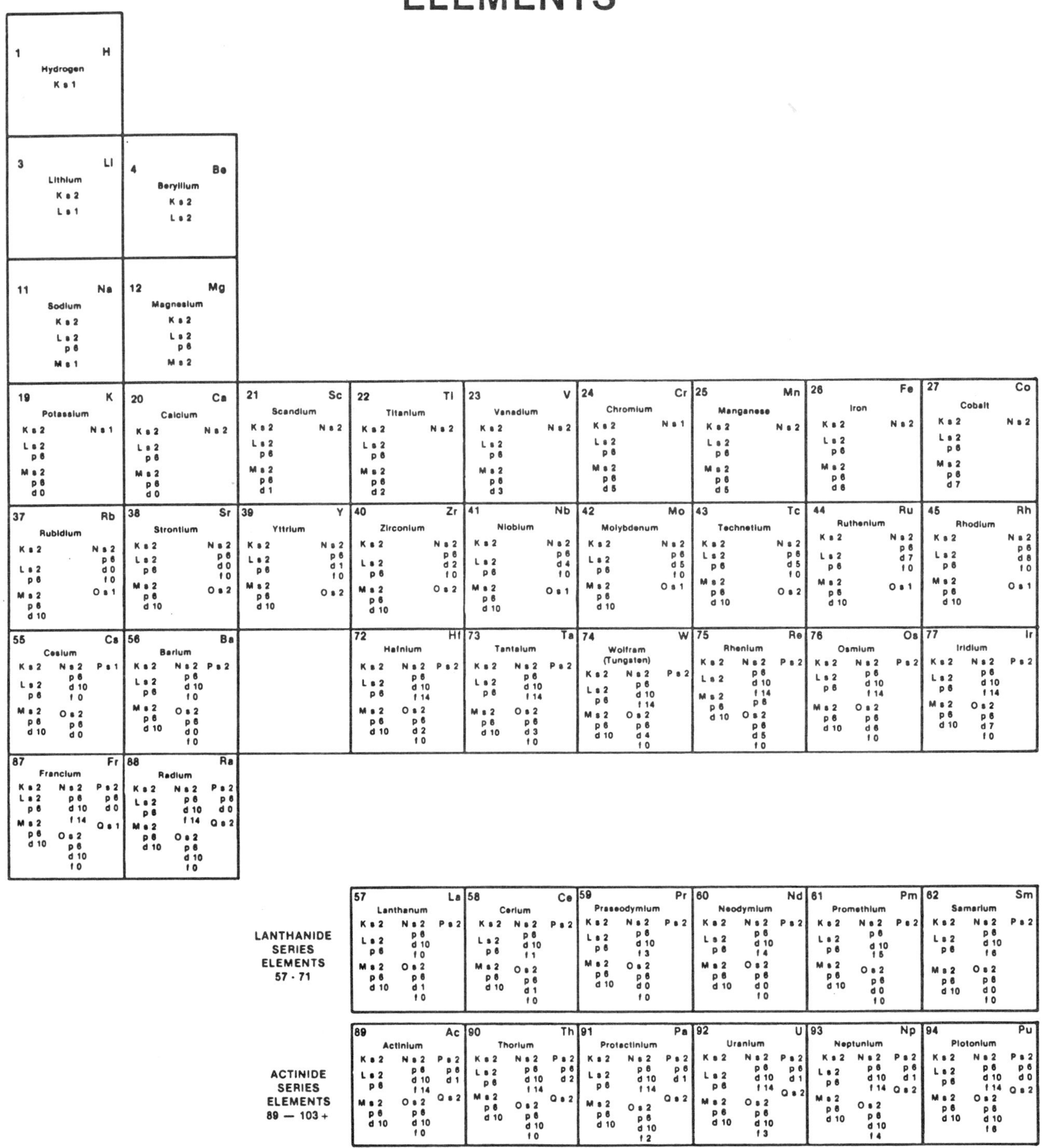

Fig. 1-5 Periodic Table of Elements

PERIODIC TABLE OF ELEMENTS

No.	Symbol	Element	K	L	M	N	O	P	Q
2	He	Helium	s 2						
5	B	Boron	s 2	s 2 p 1					
6	C	Carbon	s 2	s 2 p 2					
7	N	Nitrogen	s 2	s 2 p 3					
8	O	Oxygen	s 2	s 2 p 4					
9	F	Fluorine	s 2	s 2 p 5					
10	Ne	Neon	s 2	s 2 p 6					
13	Al	Aluminum	s 2	s 2 p 6	s 2 p 1				
14	Si	Silicon	s 2	s 2 p 6	s 2 p 2				
15	P	Phosphorous	s 2	s 2 p 6	s 2 p 3				
16	S	Sulfur	s 2	s 2 p 6	s 2 p 4				
17	Cl	Chlorine	s 2	s 2 p 6	s 2 p 5				
18	Ar	Argon	s 2	s 2 p 6	s 2 p 6				
28	Ni	Nickel	s 2	s 2 p 6	s 2 p 6 d 8	s 2			
29	Cu	Copper	s 2	s 2 p 6	s 2 p 6 d 10	s 1			
30	Zn	Zinc	s 2	s 2 p 6	s 2 p 6 d 10	s 2			
31	Ga	Gallium	s 2	s 2 p 6	s 2 p 6 d 10	s 2 p 1			
32	Ge	Germanium	s 2	s 2 p 6	s 2 p 6 d 10	s 2 p 2			
33	As	Arsenic	s 2	s 2 p 6	s 2 p 6 d 10	s 2 p 3			
34	Se	Selenium	s 2	s 2 p 6	s 2 p 6 d 10	s 2 p 4			
35	Br	Bromine	s 2	s 2 p 6	s 2 p 6 d 10	s 2 p 5			
36	Kr	Krypton	s 2	s 2 p 6	s 2 p 6 d 10	s 2 p 6			
46	Pd	Palladium	s 2	s 2 p 6	s 2 p 6 d 10	s 2 p 6 d 10 f 0			
47	Ag	Silver	s 2	s 2 p 6	s 2 p 6 d 10	s 2 p 6 d 10 f 0	s 1		
48	Cd	Cadmium	s 2	s 2 p 6	s 2 p 6 d 10	s 2 p 6 d 10 f 0	s 2		
49	In	Indium	s 2	s 2 p 6	s 2 p 6 d 10	s 2 p 6 d 10 f 0	s 2 p 1		
50	Sn	Tin	s 2	s 2 p 6	s 2 p 6 d 10	s 2 p 6 d 10 f 0	s 2 p 2		
51	Sb	Antimony	s 2	s 2 p 6	s 2 p 6 d 10	s 2 p 6 d 10 f 0	s 2 p 3		
52	Te	Tellurium	s 2	s 2 p 6	s 2 p 6 d 10	s 2 p 6 d 10 f 0	s 2 p 4		
53	I	Iodine	s 2	s 2 p 6	s 2 p 6 d 10	s 2 p 6 d 10 f 0	s 2 p 5		
54	Xe	Xenon	s 2	s 2 p 6	s 2 p 6 d 10	s 2 p 6 d 10 f 0	s 2 p 6		
78	Pt	Platinum	s 2	s 2 p 6	s 2 p 6 d 10	s 2 p 6 d 10 f 14	s 2 p 6 d 9 f 0	s 1	
79	Au	Gold	s 2	s 2	s 2 p 6 d 10	s 2 p 6 d 10 f 14	s 2 p 6 d 10 f 0	s 1	
80	Hg	Mercury	s 2	s 2 p 6	s 2 p 6 d 10	s 2 p 6 d 10 f 14	s 2 p 6 d 10 f 0	s 2	
81	Tl	Thallium	s 2	s 2 p 6	s 2 p 6 d 10	s 2 p 6 d 10 f 14	s 2 p 6 d 10 f 0	s 2 p 1	
82	Pb	Lead	s 2	s 2 p 6	s 2 p 6 d 10	s 2 p 6 d 10 f 14	s 2 p 6 d 10 f 0	s 2 p 2	
83	Bi	Bismuth	s 2	s 2 p 6	s 2 p 6 d 10	s 2 p 6 d 10 f 14	s 2 p 6 d 10 f 0	s 2 p 3	
84	Po	Polonium	s 2	s 2 p 6	s 2 p 6 d 10	s 2 p 6 d 10 f 14	s 2 p 6 d 10 f 0	s 2 p 4	
85	At	Astatine	s 2	s 2 p 6	s 2 p 6 d 10	s 2 p 6 d 10 f 14	s 2 p 6 d 10 f 0	s 2 p 5	
86	Rn	Radon	s 2	s 2 p 6	s 2 p 6 d 10	s 2 p 6 d 10 f 14	s 2 p 6 d 10 f 0	s 2 p 6	

No.	Symbol	Element	K	L	M	N	O	P	Q
63	Eu	Europium	s 2	s 2	s 2 p 6 d 10	s 2 p 6 d 10 f 7	s 2 p 6 d 0 f 0	s 2	
64	Gd	Gadolinium	s 2	s 2	s 2 p 6 d 10	s 2 p 6 d 10 f 7	s 2 p 6 d 1 f 0	s 2	
65	Tb	Terbium	s 2	s 2	s 2 p 6 d 10	s 2 p 6 d 10 f 8	s 2 p 6 d 1 f 0	s 2	
66	Dy	Dysprosium	s 2	s 2	s 2 p 6 d 10	s 2 p 6 d 10 f 10	s 2 p 6 d 0 f 0	s 2	
67	Ho	Holmium	s 2	s 2	s 2 p 6 d 10	s 2 p 6 d 10 f 11	s 2 p 6 d 0 f 0	s 2	
68	Er	Erbium	s 2	s 2 p 6	s 2 p 6 d 10	s 2 p 6 d 10 f 12	s 2 p 6 d 0 f 0	s 2	
69	Tm	Thulium	s 2	s 2	s 2 p 6 d 10	s 2 p 6 d 10 f 13	s 2 p 6 d 0 f 0	s 2	
70	Yb	Ytterbium	s 2	s 2 p 6	s 2 p 6 d 10	s 2 p 6 d 10 f 14	s 2 p 6 d 0 f 0	s 2	
71	Lu	Lutetium	s 2	s 2	s 2 p 6 d 10	s 2 p 6 d 10 f 14	s 2 p 6 d 1 f 0	s 2	
95	Am	Americium	s 2	s 2 p 6	s 2 p 6 d 10	s 2 p 6 d 10 f 14	s 2 p 6 d 10 f 7	s 2 p 6 d 0	s 2
96	Cm	Curium	s 2	s 2 p 6	s 2 p 6 d 10	s 2 p 6 d10 f 14	s 2 p 6 d 10 f 7	s 2 p 6 d 1	s 2
97	Bk	Berkelium	s 2	s 2 p 6	s 2 p 6 d 10	s 2 p 6 d 10 f 14	s 2 p 6 d 10 f 9	s 2 p 6 d 0	s 2
98	Cf	Californium	s 2	s 2 p 6	s 2 p 6 d 10	s 2 p 6 d 10 f 14	s 2 p 6 d 10 f 10	s 2 p 6 d 0	s 2
99	Es	Einsteinium	s 2	s 2 p 6	s 2 p 6 d 10	s 2 p 6 d 10 f 14	s 2 p 6 d 10 f 11	s 2 p 6 d 0	s 2
100	Fm	Fermium	s 2	s 2 p 6	s 2 p 6 d 10	s 2 p 6 d 10 f 14	s 2 p 6 d 10 f 12	s 2 p 6 d 0	s 2
101	Md	Mendelevium	s 2	s 2 p 6	s 2 p 6 d 10	s 2 p 6 d 10 f 14	s 2 p 6 d 10 f 13	s 2 p 6 d 0	s 2
102	No	Nobelium	s 2	s 2 p 6	s 2 p 6 d 10	s 2 p 6 d 10 f 14	s 2 p 6 d 10 f 14	s 2 p 6 d 0	s 2
103	Lw	Lawrencium	s 2	s 2 p 6	s 2 p 6 d 10	s 2 p 6 d 10 f 14	s 2 p 6 d 10 f 14	s 2 p 6 d 1	s 2

Fig. 1-5 Periodic Table of Elements continued

energy an electron possesses, and that energy, on the atomic level, comes in packages of a specific size. Such an energy package is called a *quantum*.

If an electron is struck by a quantum, the electron will absorb the energy and jump to a path farther away from the nucleus. Eventually, this electron will give up its package of energy and will jump to an energy level closer to the nucleus. The liberated quantum can then be absorbed by another electron, starting the cycle all over again.

In certain cases, particularly among the closely packed atoms of a metallic substance, an electron may absorb enough energy to move to a position so far from its nucleus that the attraction of the opposite charges will no longer hold it in place. It is then said to have moved into a *conduction level* or *conduction band*, becoming a *free* or *conduction electron*. We will return to this case when we discuss the behavior of electrical charge carriers in the next section.

D. The Behavior of Electrical Charge Carriers in Substances

Materials may exhibit one of three possible physical states — gas, liquid, or solid.

The atoms of a pure gas usually form molecules, but the atoms or molecules of a gas tend to be rather far apart and are free to drift around. All the electrons remain in their stable energy levels, when the gas is free of external influences.

For our example, let us consider a clear glass pipe filled with some gas such as neon. In the pipe there are also two pieces of metal called *electrodes* to which we can add or take away electrons. Adding a number of electrons to one of the metal electrodes and removing an equal number from the other creates an electric field within the tube (Fig. 1-6).

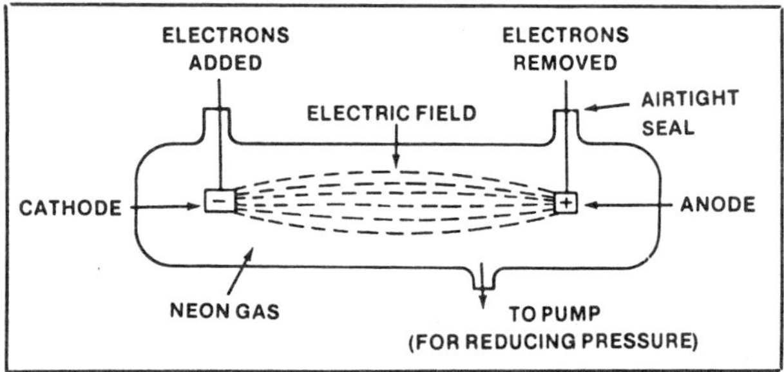

Fig. 1-6 A glow discharge tube used for examining the behavior of the charge carriers in a gas.

In general, since most of the molecules of the gas have no net electric charge, and if the gas is at atmospheric pressure and room temperature, the electric field will have little or no effect. The gas acts as an insulator, since there are very few charged bodies present that can be affected by the electric field between the electrodes.

Let us reduce the pressure of the gas in the tube and gradually increase the strength of the electric field by adding more electrons to one electrode and removing more of them from the other. Suddenly, when the electric field reaches a particular strength (depending on the nature of the gas, its pressure within the tube, and the distance between the electrodes) the gas will begin to glow. Many of the additional electrons that we added to give the negative electrode its charge will seem to disappear from it and transfer themselves to the positively charged electrode.

As pointed out earlier, we can consider this movement of electrons from the negative to the positive electrode as an *electric current* flowing from the positive electrode to the negative one. How is this possible and why did it suddenly begin to occur when the electric field reached a particular strength? The answer is: when the electric field reached a certain strength, the energy in the field was absorbed by some of the outermost electrons in the atoms of the gas, freeing these electrons from their atoms and raising them to a conduction level. This produced an equal number of free, negatively charged electrons and positively charged, ionized atoms within the gas mixture. The positively charged ions began to move toward the negatively charged electrode because of the attractive force between the unlike charges. At the same time, the electrons began to move toward the positively charged electrode.

When an ionized gas atom or molecule reaches the negatively charged electrode, it picks up an electron and returns to its normal, uncharged state. An electron which reaches the positively charged electrode, will fill up the deficiency.

At the end of the process, we again have non-ionized gas molecules or atoms, fewer electrons at the negative electrode, and additional electrons at the positive electrode. If there is still a fairly strong electric field in the tube of gas and if there are still free electrons and ions among the uncharged gas molecules, this process will continue, until all the excess electrons in the negative electrode have been removed and all the deficiency of the positive electrode has been made up.

At this point, the electric field will disappear and the process will stop. The glow that we see in the tube if the gas used happens to be neon, is caused by the collision of conduction electrons and ions within the column of gas.

All gases produce a glow, but only some of them, such as the neon and krypton used in the familiar advertising signs, produce a glow that is directly visible to the human eye. When a collision occurs, an electron will give up the energy that it absorbed from the electric field and fall back into a valence energy level around the atom. The energy given up is in the form of light, and is equal to the amount of energy originally absorbed by the electron when it moved from the valence to the conduction energy level. Once it has returned to the valence level, the electron may again absorb a packet of energy from the electric field and return to the conductance band, starting the whole process over again.

Such tubes of gas are called glow discharge tubes. Neon tubes and bulbs are used in advertising signs and as power indicator lamps in electronic equipment. Fluorescent lamps are discharge tubes whose inner walls are covered with a material that absorbs the ultraviolet glow of mercury vapor and re-emits it as visible light. Discharge tubes are also used in avionics equipment to prevent damage due to nearby lightning strikes.

For example, when lightning strikes near the cables connecting a control tower with other parts of the airport, the energy transferred to the cables ionizes a bit of neon gas in a protective discharge tube. The charge on the cables is then transferred to the ground, protecting any sensitive electronic equipment which might be attached to the cables.

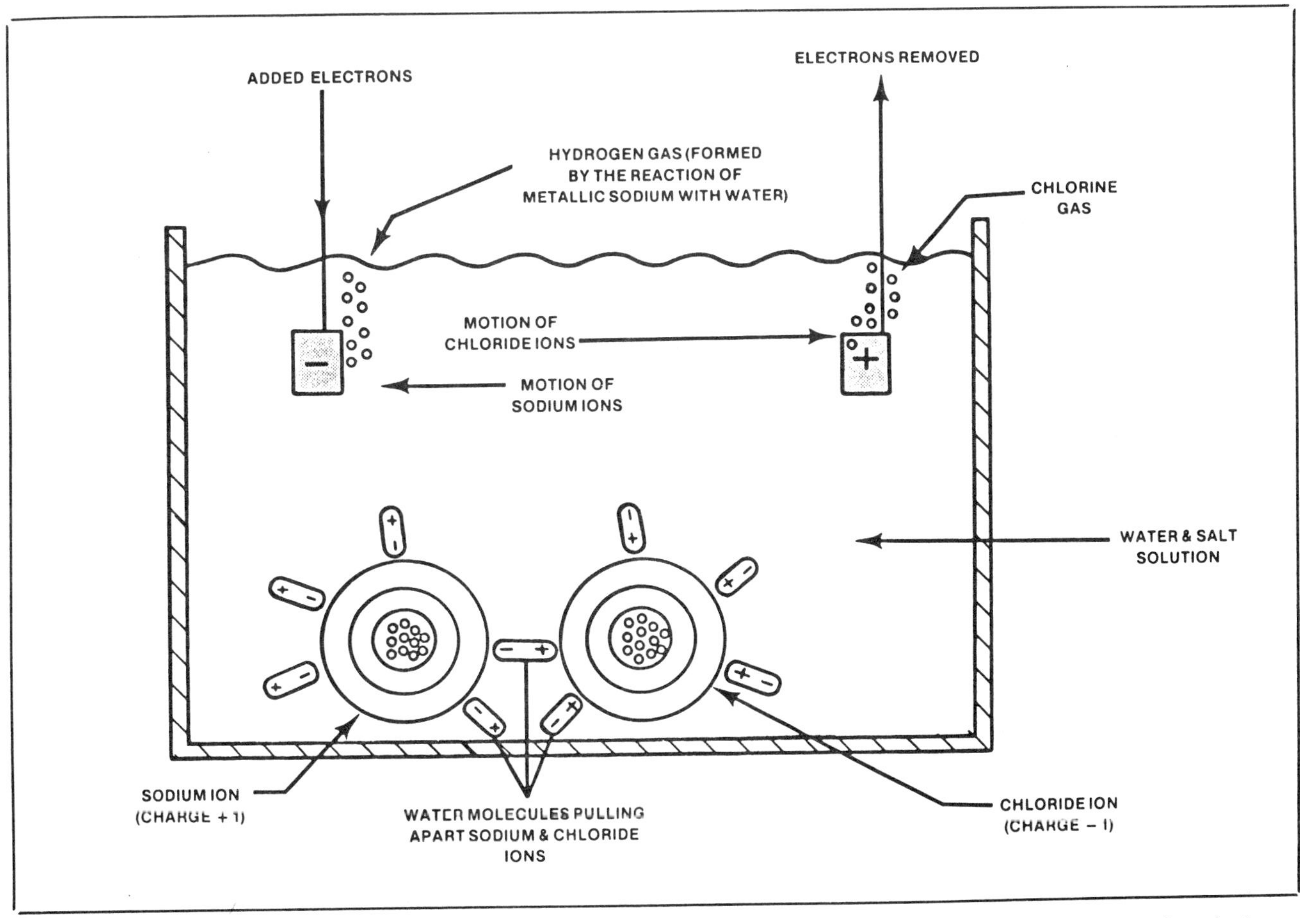

Fig. 1-7 Ionic conduction in a water solution. Notice how the ionic bond of the salt molecule is broken by the water molecules.

In liquids, a process occurs which is different from the ion and conduction electron process in an atomic or covalent bond gas described above. The fact that the atoms of a liquid are packed together much more closely than in a gas makes it unlikely that a conduction electron would get far before being picked up by an atom, forming a negatively charged ion. Since most of the liquids with which we are familiar consist of molecules of various materials dissolved in water, these dissolved substances are ionized without the presence of an electric field. This is due to what are called the *dielectric properties* of the water molecule.

Unlike most ionic molecules, in which the positive and negative ions act as though they occupied the same point in space, each water molecule acts as though it were shaped like a tiny rod, positively charged at one end and negatively charged at the other. Thus, although the water molecule is, as a whole, electrically uncharged, it is capable of electrically pulling apart the molecules of many electrovalent and even some covalent compounds.

Consider ordinary table salt, sodium chloride, which is composed of one positive sodium ion associated with one negative chloride ion per molecule. Salt dissolves in water to form positive sodium ions and negative chloride ions because of the polarity of the water molecules. If positive and negative electrodes are inserted into a salt and water solution, as shown in Fig. 1-7, the ions will begin to move toward one of the electrodes under the influence of the electric field.

Ions move rather slowly among the densely packed molecules of a liquid, more slowly than they do in a gas under low pressure, and much more slowly than electrons in an ionized gas (recall that an electron weighs 1850 times less than a single proton). When the negatively charged chloride ions reach the positive electrode, they give up their excess electron and combine in pairs to form chlorine gas bubbles. Sodium ions reaching the negative electrode accept an electron and become sodium atoms: the sodium reacts immediately with a water molecule and releases hydrogen, which forms bubbles.

As in the case of a gas, electrons are transferred under the influence of an electric field, but unlike the gas discharge tube, all the charge carriers in a water-based solution are ions. When all the added ions have been used up, the transfer of electrons stops, for water does not ionize sufficiently to sustain it.

The process of charge carrier movement in solids is somewhat more complicated than in gases or liquids, but only because of the greater variety of structural possibilities exhibited by elements and compounds in the solid state.

Amorphous or *non-crystalline* solids are those in which the closely-packed atoms of the substance exhibit no regular arrangement in space. Substances like glass and plastics fall into this category. Their behavior is rather like that of liquids, except that the atoms or molecules and ions of a non-crystalline solid are not so free to move about as ions of a liquid. (In spite of this, some non-crystalline solids do exhibit flowing over a very long time, as examination of very old window panes will reveal.)

In general, non-crystalline solids do not permit the passage of charge carriers and, like glass, are insulators. They are valuable in electronic applications by virtue of their opposition to the flow of electric currents.

The majority of the materials with which we deal in electronics form *crystalline solids*. In a crystal, the atoms are arranged in a precise, repeating geometric pattern and tend to be locked in place by the bonds formed with neighboring atoms.

In the basic crystal pattern exhibited by most of the metals used in the electronics industry to conduct electricity (such as silver, gold, and copper), we find fourteen atoms arranged in a cube with a minimum of wasted space. This pattern, called the *face-centered cubic structure*, is represented in Fig. 1-8.

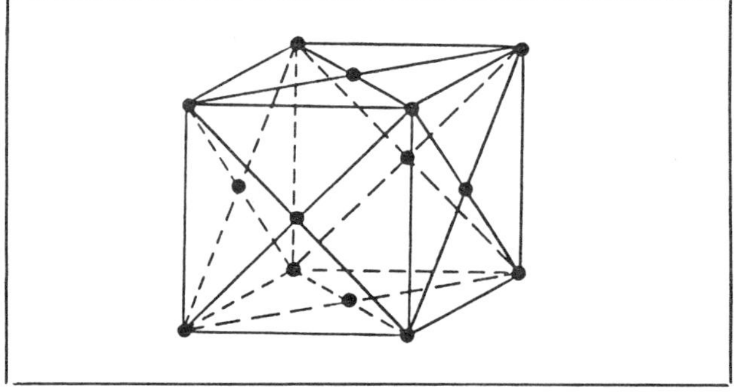

Fig. 1-8 Face-centered cubic crystal structure (one unit) — many good conductors of electricity such as gold, silver, and copper show this crystal structure.

Unfortunately, this standard representation does not give a very good idea of either the structure or closeness of packing in the face-centered cubic structure. A much better impression will be obtained if you actually construct a model, following the diagram, as suggested in the experiments at the end of this section.

Because of the overlapping of the electron shells of neighboring atoms in the face-centered cubic arrangement, we cannot say that an outer electron *belongs* to any specific atom. The valence and conduction energy levels of all the atoms in the substance coincide, and any valence electron may be thought of as belonging to *all* the atoms of the mass. Thus, although the positively charged nucleii and their inner electrons are rather firmly fixed in what is called the *crystal lattice*, the conduction/valence electrons must be thought of as a sort of cloud equally distributed through the material.

Unlike the previously discussed example of the gas, which required energy from the electric field to form conduction electron-ion pairs, in metals such as copper, which are classified as *conductors*, electrons are already in the conduction band. In fact, the addition of too much energy, in the form of heat, for example, will actually impede the passage of electrons through the crystal lattice.

Although it is easier to initiate the flow of electric current in a metallic conductor crystal lattice than it is in either an ionic solution or in a gas, not all metallic conductors provide as easy a path for the electrons as do copper, silver, or gold. This is most easily understood by considering the amount of overlap between the conduction level and the valence level as varying with the type of atom and the type of crystal structure involved. Lead, for example, forms the same face-centered crystal lattice as copper but does not offer nearly so good a path for the flow of electrons.

In nonmetals, such as sulfur, none, or very few, of the outer electrons are in a conduction band; the material does not provide a path for charge carriers and is classified as an insulator.

Midway between the conductors and insulators is a class of crystalline materials known as *semiconductors*. The most important of these are germanium (atomic number 32) and silicon (atomic number 14). Notice that these elements fall into

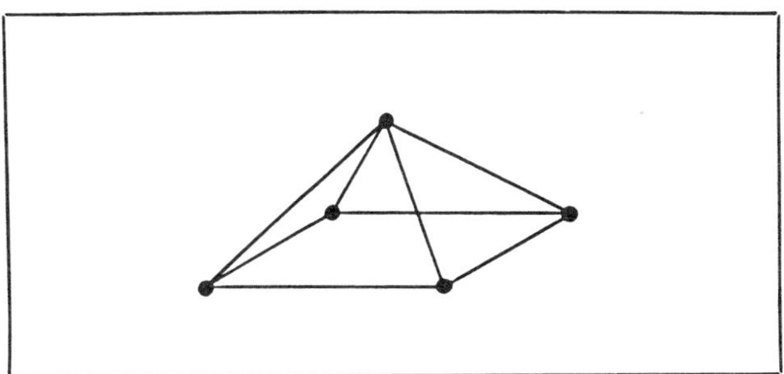

Fig. 1-9 Diamond crystal structure — the crystal structure of semiconductors.

the same column in the periodic table, and possess similar chemical and physical properties. Both form crystals whose basic unit is called the *diamond structure*, pictured in Fig. 1-9. In the diamond structure crystal each atom forms a covalent bond, sharing a pair of electrons with each of four neighboring atoms. At low temperatures, all the valence electrons are bound up in covalent bonds and the material acts as an insulator. However, when energy is added, the atoms in the lattice begin to vibrate. If this vibration is strong enough, valence electrons will absorb enough energy to enter a conduction level. The freed electrons then become conduction electrons and are free to move within the crystal under the influence of an electric field.

When an electron leaves a covalent bond between two semiconductor atoms, the broken bond behaves exactly as though it were a new particle with a positive charge free to move in the crystal. Of course the broken bond is not a real particle, but it behaves exactly as if it were, and it is called a *hole*.

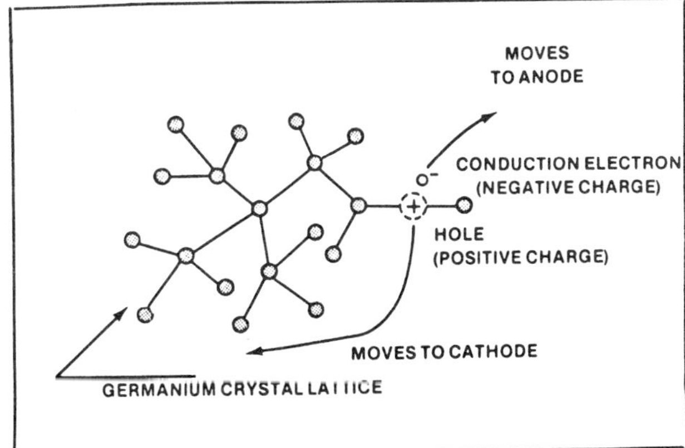

Fig. 1-10 Conductor behavior in a pure semiconductor crystal. When the electron absorbs sufficient energy, it leaves the atom, creating an electron/hole pair.

If a rod of silicon crystal is placed in an electric field and sufficient energy is added to it to cause the atoms in the crystal lattice to vibrate, we will find a migration of conduction electrons toward the positive electrode and holes toward the negative electrode (Fig. 1-10).

A. Self-Test Questions:

Complete the following statements:

1. The negatively charged particle found in the atom is called ____________.
2. The ________ has a positive charge and is found in the nucleus of an atom.
3. An uncharged particle found in the nucleus of an atom is the ________; it serves ________.
4. An outer electron which an atom may lose or gain in the process of forming a compound is called a ________ electron.
5. An electrically charged body is surrounded by an ____________.
6. When an atom gives up an electron to another atom it becomes an ________. When an atom shares a pair of electrons with another atom, it forms a ________ bond.
7. An electrode to which electrons have been added is called a ________; it has a ________ charge.
8. Conduction of electricity in a gas takes place through the movement of ________ and ________ under the influence of an ________ ________.
9. Conduction of electricity in a semiconductor takes place through the movement of ________ and ________ under the influence of an ________ ________.
10. A quantum is a minimal ________ of ________.

Briefly answer the following:

11. How can absorption of energy affect an electron?
12. What is an element? How many are there?
13. What is an electron shell?
14. What is the difference between a valence electron and a conduction electron? In what materials do we often find conduction electrons?
15. What is the crystalline structure of copper? Of germanium?
16. What are the principal differences among conductors, insulators, and semiconductors?

B. Questions For Thought:

1. We are all aware that much of the energy we use can be traced back to sunlight. Coal, oil, and natural gas all started this way. Consider the following devices; from what can their ability to do work be traced?
   a. A horse
   b. An atomic reactor
   c. A windmill
   d. A volcano
   e. A waterfall
2. In what ways can knowledge of the composition of an atomic nucleus help predict whether a particular material will provide passage for an electric current? How does the nucleus directly take part in this process?
3. Using the periodic table, Fig. 1-5, divide the following elements into groups that should possess similar chemical properties. Try to predict what sort of ions, if any, the groups would form.

Element	Atomic Number
lithium	3
radon	86
bromine	35
sulfur	16
fluorine	9
argon	18
potassium	19
tellurium	52
xenon	54
sodium	11
chlorine	17

C. Experiments and Demonstrations:

1. Construct models of the face-centered cubic and diamond crystal structures pictured in Fig. 1-8 and Fig. 1-9, using ping pong balls, cut plastic straws, and polystyrene cement. Notice that each model represents only the basic unit, which is repeated many times in a crystal.

2. Connect a 115 volt neon glow lamp (NE-2), a 220,000 ohm 1 watt resistor and a 50-200 volt DC variable voltage supply as shown below.

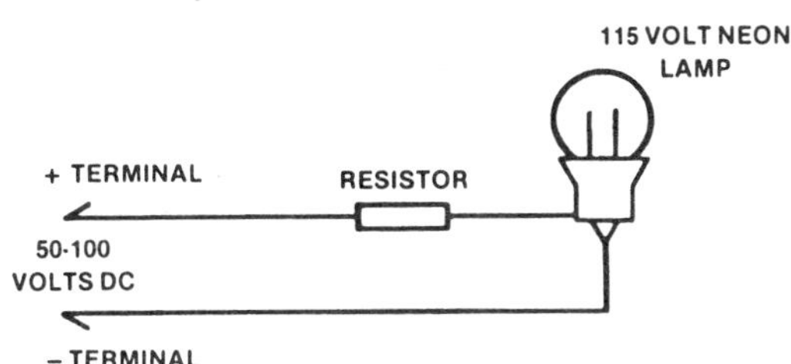

Slowly increase the voltage of the supply from 50 volts until the bulb glows. Does the glow begin gradually or all at once? At about what voltage does the lamp begin to glow? What do you think is the purpose of the resistor? *Hint:* an ionized gas is a fairly good conductor of electricity.

CAUTION:

The potentials used in this demonstration can be dangerous. Exercise caution.

SECTION II

The Electric Field and Electric Current

In the previous section, electrical effects were mainly considered from the point of view of a single charge carrier. We discussed the place of the electron in the atom, the electric field surrounding single charge carriers, and the means by which single charge carriers pass through materials. In the field of avionics, however, we deal with charge carriers in huge numbers. For example, during one second, the astronomical number of 6×10^{15} or 6,000,000,000,000,000 electrons may pass through a point in an average, low-power, solid-state circuit. Since it requires a continuous passage of twenty times this number of electrons to light a small indicator lamp, it is easy to see that we shall have to consider electrical effects in terms different from those that apply to a single charge carrier. For this reason, this section begins with a further consideration of the electric field and introduces the concepts of *potential* and *potential difference.*

The physical laws that govern electrical effects, as well as the basic units by which these effects may be measured, were the subject of research throughout the Eighteenth Century. In this section we shall trace the history of this research, showing how the basic quantities used today for the measurement of quantity of charge, strength of an electric field, and so forth, were derived. In order to provide an easier transition from the material presented in the first section, the historical distinction between static and current electricity is in part retained, although in reality, both are manifestations of the same electrical effects under differing conditions.

A. *The Electric Field*

As mentioned in Section 1, the study of avionics can be greatly simplified by keeping in mind that it is the study of some of the means of tapping and controlling energy to perform useful work.

Because of the infinitesimal size and charge of the individual electron or ion, it is necessary for the avionics technician to think in terms other than that of single charge carriers. The most likely point of departure is to consider the range of visible electrical effects in order to gain an insight into electrical phenomena. In spite of the large number of charge carriers required, it is not difficult to obtain electrical effects that are visible and measurable.

All ancient people were aware of one or more forms of static electricity. The Greeks, for example, observed that if a piece of amber (a hard gemstone formed when pine tree sap is petrified) is rubbed with cloth, the amber develops the ability to attract small bits of wood or cloth. These, after touching the amber, would fly off again.

In our scientific age, the reason for this is widely known: contact between the cloth and the amber (contrary to popular belief, contact alone is sufficient, the rubbing merely strengthens the effect) causes the transfer of a fairly large number of electrons from the cloth to the amber. Since both the cloth and the amber began with an equal balance of negative electrons and positive protons in their structure, the transfer of some electrons to the amber results in a negative charge — an excess of negative charge carriers — in the amber, causing a deficiency of electrons — an equal positive charge — on the cloth.

Thus far, the mechanism is rather simple and widely understood. But why should a charged bit of amber be able to affect electrically neutral objects some distance away? Traditionally, this has been one of the most mysterious of electrical effects; the fact that an electrically charged body, even a single charge carrier, is capable of acting over a distance, exerting an attracting or repelling force.

As late as the early twentieth century, many scientists refused to accept the idea of action over a distance, and they actually believed in something they called the *electromagnetic ether,* which was thought to fill all space and to provide a means for transmitting light as well as magnetism and the forces of electrical attraction and repulsion. We now know that the electromagnetic ether does not exist. In fact, an electrical charge changes the nature of the space around it in such a way as to set up an electric field. It is this change in the nature of the space around a charged body that is responsible for the forces upon other bodies that are brought into the neighborhood.

1. *Lines of force*

During the first half of the Nineteenth Century, Michael Faraday, one of the pioneers of electrical experimentation, introduced a method of charting and visualizing electric or magnetic fields. Faraday's method was to draw a series of lines representing the electric field around a charged body in such a way as to show the direction of the force that the field would exert on a small test charge placed at various points. This enables us to picture the direction and shape of even complex electric fields caused by a number of neighboring charged bodies.

A further refinement to Faraday's method was to vary the number of lines of force drawn through an area in the field in such a way as to give some indication of the strength of the force exerted on a test charge situated in that area. In electric field diagrams, such as those in Fig. 2-1 areas of greater field strength are shown by the bunching of the lines of force. Areas of relatively weak field strength are shown by a spreading of the lines of force. The direction arrows added to the lines for force in Fig. 2-1 represent a third quantity symbolized by this method. They show the direction that a positively charged test body would move if placed in the field.

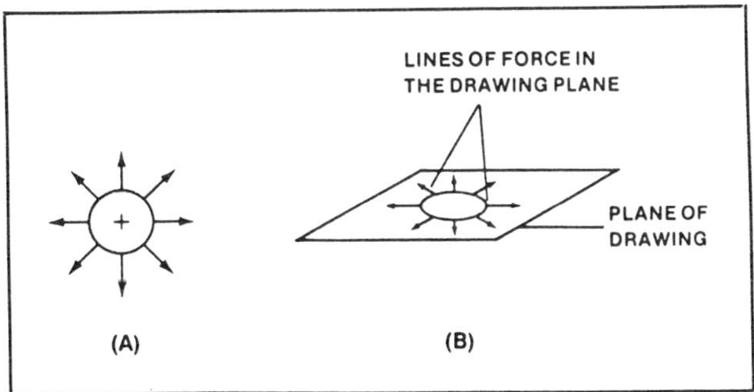

Fig. 2-1 Faraday's method of charting the electric field around charged bodies.

Using lines of force, it is possible to picture the electric field around a single charged sphere, as in Fig. 2-1A, or around two spheres with unlike (Fig. 2-2) or like charges (Fig. 2-3). It should be noted that although these figures are two dimensional, the electric field is three dimensional. Therefore, the illustrations represent the field only as it would exist on a very thin piece of paper slicing through the sphere or spheres, as shown in Fig. 2-1B.

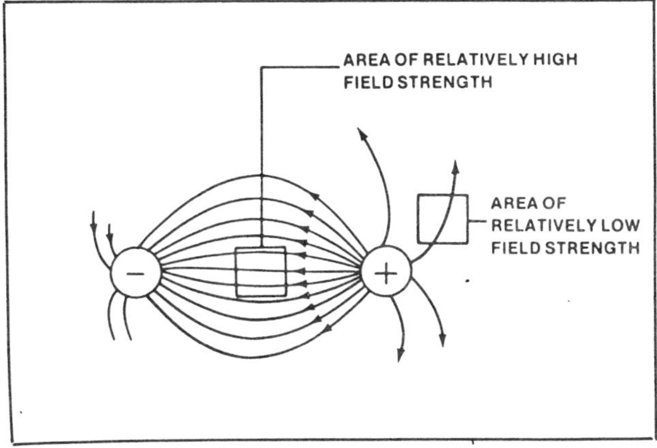

Fig. 2-2 Lines of force between bodies with unlike charges. Notice that the region of the greatest field strength is between the bodies.

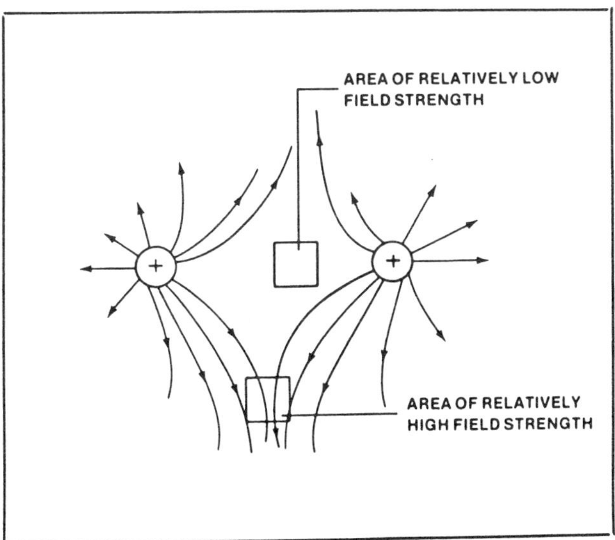

Fig. 2-3 Lines of force between bodies with like charges. The region between the bodies is relatively free of lines of force. The field is weak there.

As may be seen from the figures, the existence of two charged bodies close to one another profoundly modifies the shape of the electric field in their neighborhood. The charting of the electric field in the neighborhood of several charged bodies, or around a charged body near conducting surfaces, or around even a single charged body of irregular shape, is by no means an easy task.

Fortunately, it is possible to determine mathematically the strength and direction of an electric field at a given point. This is made possible by two facts: first, the charge on any body is found wholly on the outer surface of that body, and second, the strength and direction of the electric field at any point is equal to the sum of the effects caused by *all* the charged bodies in the neighborhood. Therefore, if we want to know the strength and direction of the field in the neighborhood of several charged bodies, we have to calculate the strength and direction caused by each separate field and add all the effects, being careful to preserve the direction of each. This sort of additon is called a *vector sum*, for each field has not only a strength, but a direction also. It should be noted that each force, $\vec{F}$, in Fig. 2-4 is marked with a little arrow to show that it is a quantity with an assigned direction (a vector quantity in mathematical terms).

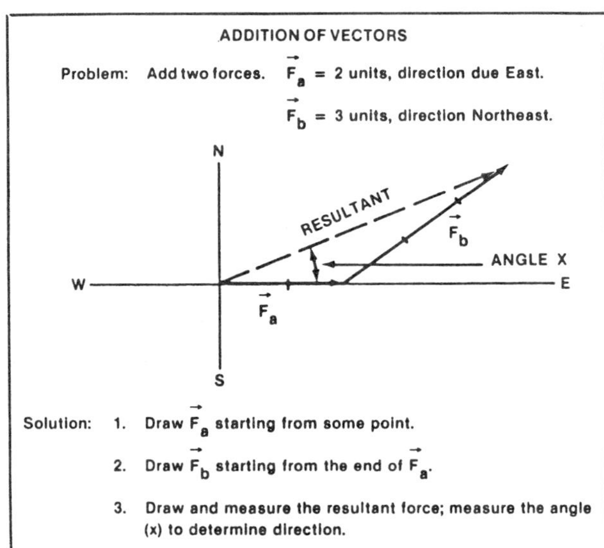

Fig. 2-4 Addition of vectors.

Vector quantities such as force cannot be added in the same way as we would add the number of apples in two sacks. For example, the result of a 50 km/hour wind from the south and a 50 km/hour wind from the east is *not* a 100 km/hour wind to the northwest, but a 70.5 km/hour wind to the northwest. Fig. 2-4 shows a typical example of a vector addition of strengths and directions.

Similarly, to calculate field strength and direction at a distance from an irregularly shaped charged body, it is necessary to divide the body into simpler shapes and calculate the field strength and direction due to each portion as though it were a separate body. The vector sum of the effects due to each section, when added vectorially, will picture the total field at that point.

In performing the experiments at the end of this section, you may notice that when a bit of styrofoam touches a charged object it will then fly away again. By now it should be clear that this happens because the styrofoam itself becomes charged with the same sort of charge as the body which attracted it. This interesting process illustrates a very important principle — the difference between the charging of an object by contact and the charging of an object by induction. In Fig. 2-5A, when the styrofoam ball approaches the charged body, the electric field from

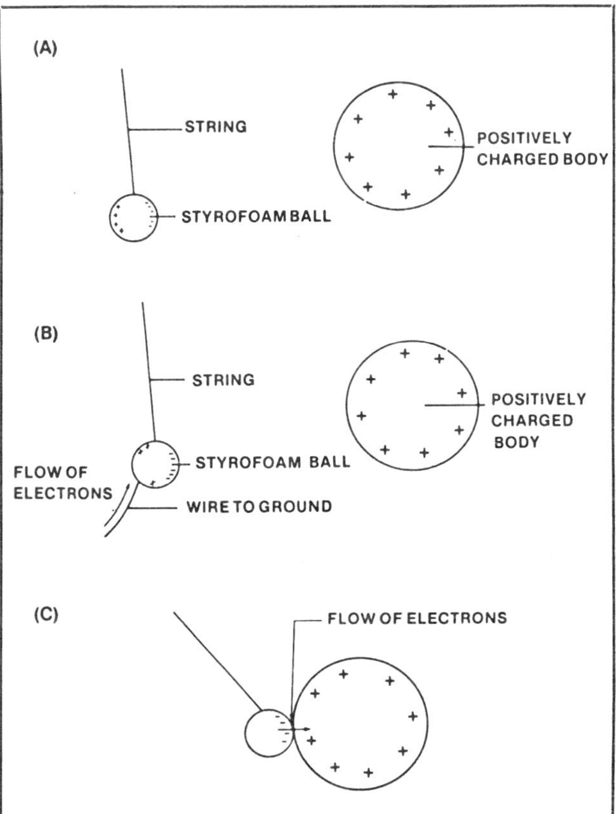

Fig. 2-5 A — In the process of inductive charging the styrofoam ball is affected by a charged object causing electrons to migrate to one side of the ball. B — Grounded wire is touched to styrofoam ball. C — Strong negative charge on the styrofoam ball pulls it into contact with positively charged object.

the charged body exerts a force upon the electrons in the styrofoam, causing some of them to drift to the surface closest to the charged body. This leaves a net positive charge on the surface of the styrofoam farthest from the charged body.

If, as shown in Fig. 2-5B, a grounded wire is touched to the surface of the styrofoam ball farthest from the charged body, electrons will flow into the styrofoam, neutralizing the net positive charge on that surface of the ball. When the grounded wire is removed, the styrofoam ball is left with a negative charge (i.e., it will be charged with a charge opposite to that of the charged body that caused the effect). This is called charging by induction. If a grounded wire is not touched to the styrofoam, and it is allowed to touch the charged body, the effect will be quite different. Some of the electrons on the surface of the styrofoam touching the charged body will drift into it. This will leave the styrofoam with a net positive charge. Since like charges repel each other, the styrofoam ball will then be pushed away from the charged body. Unlike charging by induction, charging by contact creates a *similar* charge on both objects.

2. Force and energy in the electric field

Since an electric field from a piece of rubbed amber or plastic is capable of exerting a force and lifting small objects, we see that an electric field can perform useful work. In this sense, it is very much like the gravitational field which surrounds the earth. If a weight is lifted from the earth, the system formed by the earth, the weight, and the gravitational field possesses something that it did not possess before. This something is energy, the ability to do work; for the weight may be attached to a rope and made to turn a shaft as it returns to the earth, or it may be attached to a pulley system to lift another weight. The weight is said to have a certain amount of potential energy. The amount of potential energy it possesses depends upon its relationship to the gravitational field. This is understandable since it took a certain amount of work to lift the weight against the pull of gravity. Ignoring the various losses due to friction or to the inefficiency of the lifting mechanism, the amount of work involved in lifting a weight a certain distance against the gravitational field of the earth is exactly equal to the potential energy possessed by the weight at that point. Again, neglecting the various losses that might occur because of friction or inefficiency, this is also equal to the amount of work that can be derived from letting the weight fall.

The electric field surrounding a charged body shows an effect similar to that of the gravitational field around the earth, but it is further complicated by the fact that a charged body in an electric field may be either attracted or repelled, while the forces that exist in a gravitational field are those of attraction only.

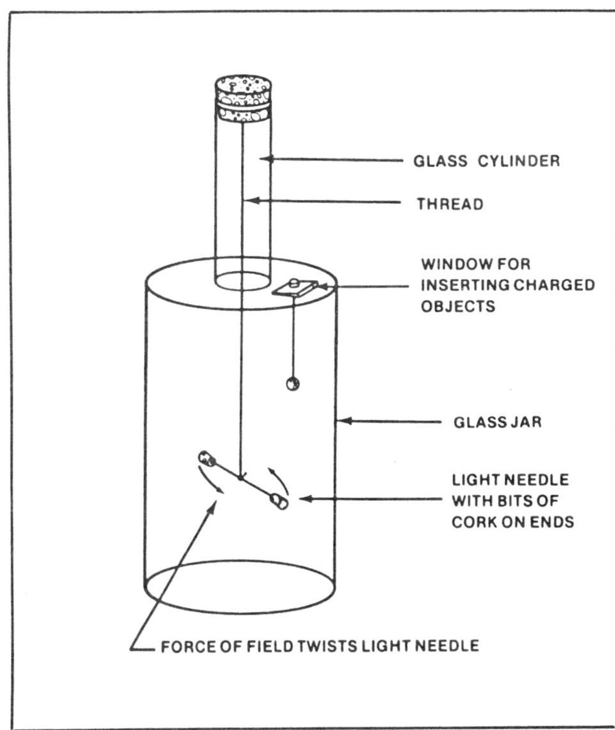

Fig. 2-6 The Coulomb Balance was used by Charles Coulomb (18th century) to study electrical forces.

Using a delicate balance whose principles of operation are illustrated in Fig. 2-6, the eighteenth-century French physicist, Charles Coulomb, was able to study the forces at work in the electric field and to define a basic unit of electrical charge in terms of the more directly measurable quantities of force and distance. After a series of experiments that would seem crude to modern scientists, Coulomb was able to deduce that the effect of the electrical field decreases as the square of the distance between two bodies increases. Using his balance and tiny magnets, he was also able to show that the forces caused by the magnetic field between two magnetized bits of iron also decrease with the square of the distance between the magnets. This relationship, known as the *inverse square law*, seems to be one of the fundamental relationships of physics. Roughly a hundred years before Coulomb's experiments, Sir Isaac Newton discovered that this relationship describes gravitational effects. Stated mathematically: the force $\vec{F}$ on either of two bodies because of a field between them is inversely proportional to the square of the distance between them, or

$$\vec{F} \sim \frac{1}{d^2}$$

where d is the distance between the bodies. Thus, if the force $\vec{F}$ is equal to one unit when the bodies are one unit apart, it will be equal to one quarter unit when the bodies are two units apart, and so on.

In addition to the distance between the two charged bodies, the force exerted by the field also depends upon the strength of the charge which gives rise to the electric field. Using this fact, Coulomb was able to define a basic unit of charge. This is the quantity which, when given to each of two bodies exactly one centimeter apart, produces a field that exerts a force of one dyne upon both bodies. (A dyne is the basic unit of force in the gram - centimeter or cgs metric system and is equal to 0.00001 newton, the basic unit of force in the mks system. In this book we will use the standard metric mks system, based on the meter, kilogram, and second as its basic units). Coulomb's findings, then, may be expressed by the mathematical equation:

$$\vec{F}\text{orce (in dynes)} = \frac{\text{charge of body A} \times \text{charge of body B}}{\text{distance (in cm)}^2}$$

or,

$$\vec{F}\text{orce} = \frac{q_A \times q_B}{d^2}$$

(Equation 2-1)

If we set the force at one dyne and the distance at one centimeter, the charge on either body will be one statcoulomb and is equal to the charge carried by 2,082,060,000 or approximately 2×10^9 electrons. In the mks system, which we will be using, the standard unit of charge is *not* the statcoulomb and is *not* based on the force between charged bodies, but rather on the force that exists between current-carrying wires.

Because of the force exerted upon it by the gravitional field, as pointed out at the beginning of this section, a weight suspended above the earth has a certain amount of potential energy. This means it has the ability to do work. Similarily, a charged body placed in the electric field produced by one or more charged bodies also has the ability to do work. Therefore, potential energy is defined as the operation of a force over a distance or:

$$PE = \vec{F} \times d. \qquad \text{(Equation 2-2)}$$

Notice that the force ($\vec{F}$) is a vector quantity while potential energy (PE) is not. By using algebra we can change equation 2-2 to:

$$\vec{F} = \frac{PE}{d}$$

This gives us a convenient way of determining the potential energy of the field at a certain distance from a charged body q_A. Recalling equation 2-1 we can define this energy in the following way:

$$\vec{F} = \frac{q_A q_B}{d^2}$$

And, since $\vec{F}$ is also equal to PE/, we can set the two equal to one another:

$$\frac{PE}{d} = \frac{q_A q_B}{d^2}$$

multiplying both sides by d we get

$$PE = \frac{q_A q_B}{d}$$

(Equation 2-3)

Where q_A and q_B are given in statcoulombs and d in centimeters, the potential energy will be measured in statvolts per coulomb. If we assume that q_B is a one statcoulomb test charge, equation 2-3 simplifies to an expression for the potential, that is, the potential energy per coulomb of test charge of the field due to q_A or

$$\text{potential (statvolts)} = \frac{q_A}{d}$$

(Equation 2-4)

We may consider equation 2-4 as the definition of the statvolt, the electrostatic unit of potential. The electrostatic units, which have been given here for purposes of introduction, are of little practical use to the avionics technician. The derivation of the so-called *practical* or mks electrical units, is reserved for later in this section.

B. The Applications of Electrostatics

For most engineers, or technicians, the effects discussed so far are of minor importance, and after working out a few exercises in school, the whole thing is quietly forgotten. It's not that simple for the avionics technician because an airplane passing rapidly through the air or through charged clouds is in a good position to pick up a very strong electric charge.

In the early days of aviation, when airplanes were made of such *insulators* as wood and fabric, the effects must have been very spectacular. We have all experienced a surprising shock after walking across a carpet and grasping a doorknob. Imagine what was in store for the hapless crewman who reached up to help a pilot out of a WWI biplane! This problem was eliminated by the use of metallic tail skids. These skids harmlessly grounded the built-up charge on those early aircraft.

With the invention of all-metal airplanes and electronic communications and navigation equipment, the problems associated with static charges arose again. One particular problem involved portions of an airplane which were insulated from each other and developed opposite charges. This often resulted in minor continuous spark discharges which caused a great deal of radio static. In time, aircraft manufacturers solved this problem by making sure that all the aircraft surfaces were connected by conducting strips, thus ensuring that they would all carry the same charge.

Use has also been made of the lightning rod principle discovered by Benjamin Franklin. He found that near a charged body of irregular shape, the electric field was strongest in the vicinity of those parts with the smallest radius of curvature. Since a sharp point has a very small radius of curvature, Franklin's lightning rod provided a means of protecting buildings from being struck by lightning. As shown in Fig. 2-7, the lightning rod consists of a pointed rod attached to the roof of a building and connected to the ground by means of a heavy wire.

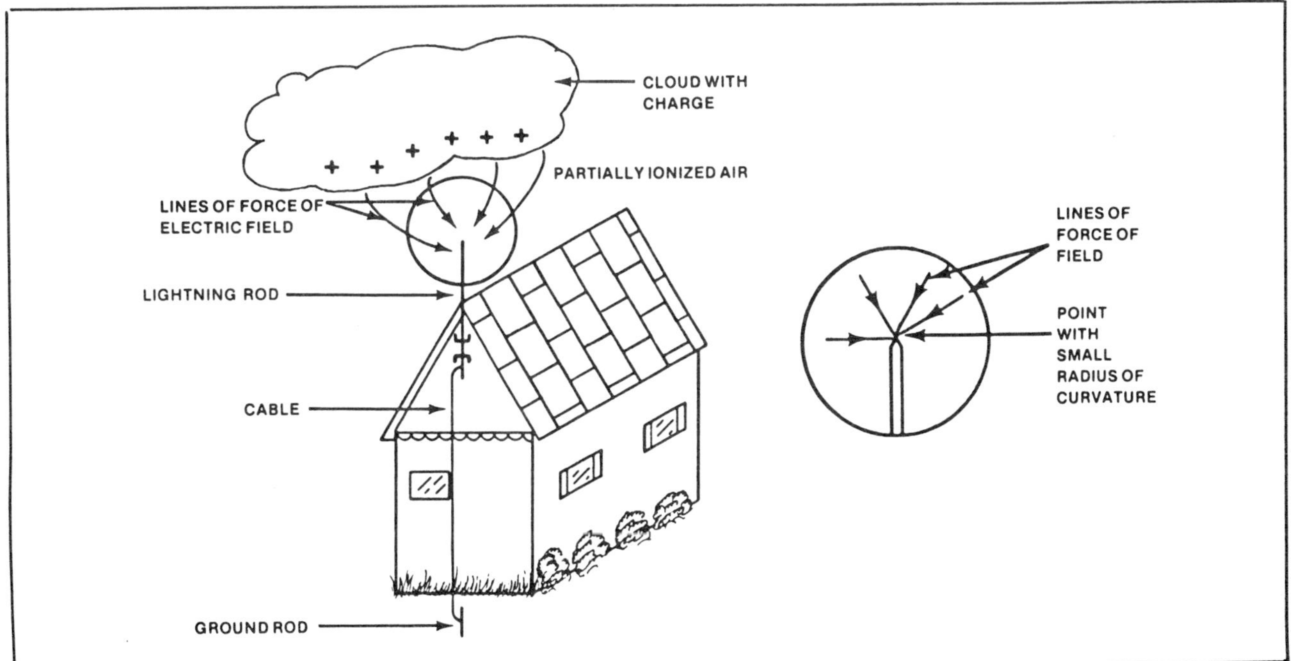

Fig. 2-7 The lightning rod was invented by Benjamin Franklin.

Since the electric field is strongest in the vicinity of the point, a region of air in the vicinity of the point will be ionized when there is a large difference in charge between the ground and passing clouds. This ionized region acts as a conductor and since it is higher than the roof of the building, potential lightning strikes will tend to be channeled toward it and into the ground via the rod and the cable, thus protecting the structure.

The principle of creating an ionized region around a sharp point is used in the modern aircraft static discharge point (Fig. 2-8). This assembly consists of a number of needle points attached to the wings of an aircraft. If the aircraft has developed a strong static electric charge opposite to that of the clouds through which it is passing, the ionized air around the needle points (assuming that these are clean and in good repair) will permit the charge to bleed off harmlessly.

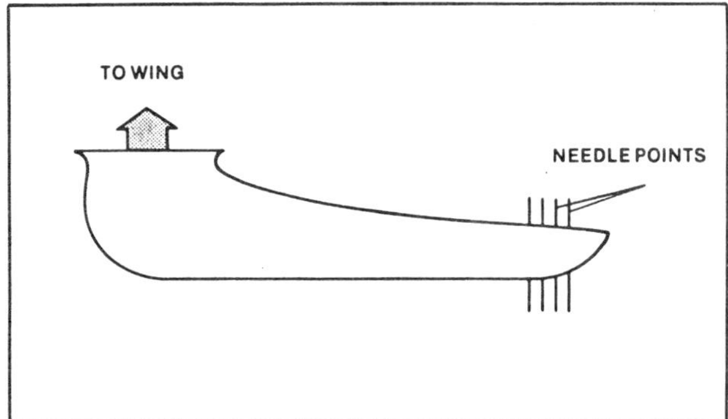

Fig. 2-8 One form of modern aircraft static discharger.

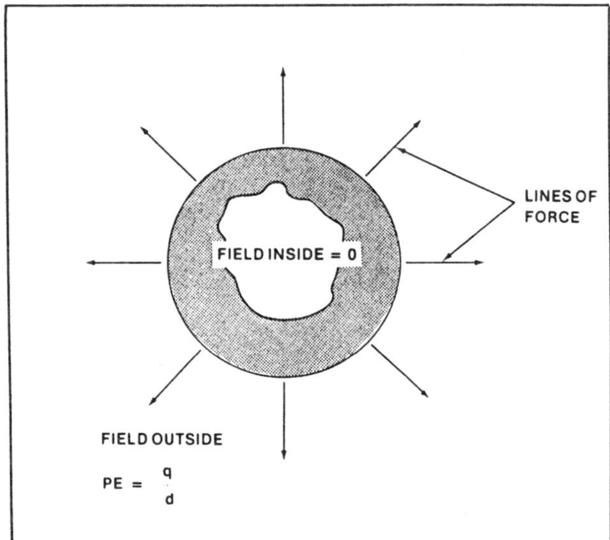

Fig. 2-9 A charged hollow sphere. Note that no field is present inside the sphere.

With the exception of the problems that might be caused to electronic equipment and radio reception, the build-up of static charges on a metal airplane has no effect at all upon the passengers inside, even though the total charge on the airplane may be very great. This is because of the curious fact that the electric field *inside* a hollow charged body is zero, that is, all of the charge appears on the outside of the body (Fig. 2-9).

C. Electric Current and the Invention of the Battery

The invention of the Leyden jar in 1754 by a group of Dutch scientists at the University of Leyden made it possible to collect large static charges. The Leyden jar (Fig. 2-10) consists of a simple glass jar lined inside and out with a layer of thin silver foil. The outer foil is connected to ground, the inner foil to a conducting ball mounted on the stopper at the top.

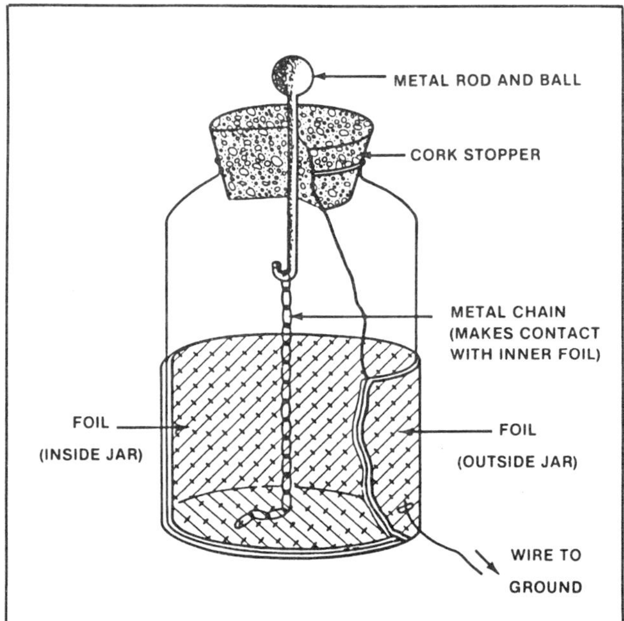

Fig. 2-10 The Leyden Jar — an early device used to collect large static charges.

If charged bodies are successively touched to the ball, the charge will be transferred to the inner foil and, by induction, an opposite charge will be transferred to the ground. Depending on the humidity of the surrounding air (moist air allows some of the charge to leak off) and the insulating qualities of the glass, quite a large charge could be stored, even in the earliest models built. This storage device permitted early experimenters to perform all sorts of bizarre demonstrations, few of which were of any use in advancing man's knowledge or ability to utilize electrical energy.

A number of French experimenters did attempt to transmit the discharge of a Leyden jar along wires through successively longer distances, and, had they possessed a suitable detector, might have even devised a primitive telegraph. For the most part, these early experiments were little more than such curiosities as attempting to cure rheumatism or kill chickens with electric shocks. The reason for this lack of practicality of the early work in electrostatics is twofold: first, static electric charges are difficult to control, and second, it is most difficult to sustain a static electric field which is actually doing some sort of work.

In our earlier discussion of the potential energy of a charged body in an electric field, it was pointed out that this potential energy is like the potential energy stored in the gravitational field when a weight is suspended above the earth. When the weight is released, this potential energy is transformed into work. The system consisting of the earth and the weight then no longer has any energy. Similiarly, if we rub a glass rod with a silk cloth, electrons are transferred from the rod to the cloth, creating an electric field between the two charged bodies. If we allow this field to do work in attracting a corner of the cloth to the rod, the electrons will re-enter the rod in the process, and the electric field will disappear. A bit of thought will show that any means of deriving work from the electric field surrounding a charged body will serve to diminish the potential of the body. Since the creation of static charges is, at the present state of technology, an inefficient process, it is not usual to find static electric effects in use where it is necessary to harness large amounts of energy.

In order to provide the easily derived, controllable energy required by modern avionics, a source of a continuous electric field is necessary, one that would not disappear when made to do a moderate amount of work in a controlled manner for a period of time. The first such source of a sustained electric field was discovered, although not recognized for what it was, by the Italian scientist Galvani in 1786. Galvani's friend, the physicist Alessandro Volta, quickly recognized that the effect his friend had noticed while experimenting with frogs' legs hanging from copper hooks on an iron grating had nothing to do with biological electricity as Galvani had thought. Instead, it was associated with the chemical effect of different metals suspended in an ionized liquid. Volta, giving full credit to his friend, called the effect *Galvanism,* and pushed his research to surprising lengths in a relatively short time.

Volta's research led to his invention of the first battery. This unit, called the *voltaic pile* or *voltaic battery* (Fig. 2-11), was comprised of a stack of alternating disks of two different metals separated by spacers made of leather which had been soaked in an ionized solution.

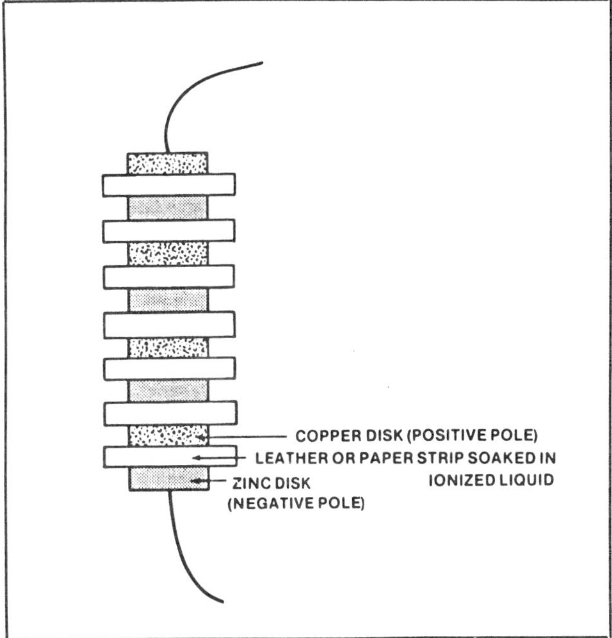

Fig. 2-11 The Voltaic Pile — the earliest documented electrochemical battery.

Although it may not be immediately evident, the voltaic battery operated rather like our experiment of rubbing a glass rod with a silk cloth. It should be recalled that rubbing is not necessary to transfer electrons from the glass rod to the silk cloth. The rubbing merely aids the transfer of electrons by bringing the glass rod and cloth into better contact and releasing some of the loosely held electrons.

Because of the nature of the materials, the electrons in the silk cloth are at a lower energy level than those in the glass rod. If the two are brought into close contact, a certain number of electrons will seek this lower energy level and enter the cloth. This process will continue until the repelling electric field caused by the accumulated excess electrons in the cloth exactly balances the tendency of further electrons to leave the glass and enter the silk.

In the voltaic battery, an ionized (and therefore tions, the wires are forced together. In either, rubbing. When two different metals are bathed by this ionized liquid, the metal whose free electrons are at a higher energy level, will give up electrons to the liquid. At the same time, the other metal, which has electrons at a lower energy level, will draw electrons from the liquid.

This exchange process is actually more complex than explained here. In general, one of the metals is gradually eaten away to furnish the ions that act as the charge carriers. As in the case of the glass rod and silk cloth, the exchange process will continue until the accumulated difference in electrical charge between the two metals just balances the energy difference of the electrons in the two metals.

At this point, we can see that since the two pieces of metal possess different charges, there will be an electric field between them. A small test charge will show the presence of this field as it is moved from the neighborhood of one piece of metal to the other.

The amount of energy that may be derived from this difference in charge in a single unit (or cell) of a voltaic battery is measured by the difference in the potential energies of a test charge at either piece of metal, commonly called the *poles* of the cell. In keeping with Franklin's belief that current comes from the positive pole of the cell (the piece of metal which has given electrons to the liquid), this potential difference is measured by using a positive test charge. At the positive pole of the cell, this positive test charge is subjected to a force resulting from the electric field between the poles, and may be said to have a relatively high potential energy.

The test charge, at this point, can be made to do work simply by allowing it to move toward the negative pole, where it will be neutralized by picking up electrons. At the negative pole, the positive test charge has no potential energy.

In general, the potential difference between the positive and negative poles usually ranges between 0.004 and 0.007 statvolt — depending upon the metals and ionized liquid being used. While this small potential difference is not capable of doing much work, it is possible to stack many such cells in a battery (Fig. 2-11) to increase this potential, as Volta discovered. When this is done, the potential difference between the two endmost poles will be equal to the sum of the potentials from all the cells. For example, stacking ninety-five cells, each of which produces a potential difference of 0.004 statvolt will produce an overall potential difference of:

N (number of cells) $\times$ P (potential of one cell)
= Total potential difference = 0.383 statvolt

which is about equal to the potential difference of the 115-volt household distribution line.

Thus far, we have only hinted at the difference between a voltaic cell and a glass rod which has been rubbed with a silk cloth. Let us take a closer look at the cell, which functions like an *electron pump*. As stated previously, the difference between potential energy levels in the two pole pieces of different metals, plus the action of the ionized liquid separating them, enables a number of electrons to migrate from the positive pole to the negative pole. This process continues until the cell develops a sufficiently large negative charge on the negative pole, and positive charge on the positive pole to offset the potential difference possessed by electrons in the two metals. The cell is now in equilibrium, and theoretically, no further electro-chemical change will occur. Practically, as we have learned from sad experience before the days of leak-proof cells, some electro-chemical action does occur, but at a very slow rate.

Now, what would happen if a piece of copper wire were connected between the pole pieces of the cell? As mentioned earlier, the presence of this conductor would change the shape of the electric field between the poles. It would be found that the electric field would lie wholly within the copper wire (Fig. 2-12). The electric field would exert a force upon the free (or conduction electrons) in the electron cloud of the copper. These electrons are in constant motion at room temperature, and, the presence of an electric field will cause a gradual net drift of electrons from the negative pole (where there is a surplus of electrons) to the positive pole (where there is a deficiency). At the same time, the electrochemical process within the cell strives to maintain the equilibrium condition previously described. The result is that for every electron that enters the positive pole of the cell

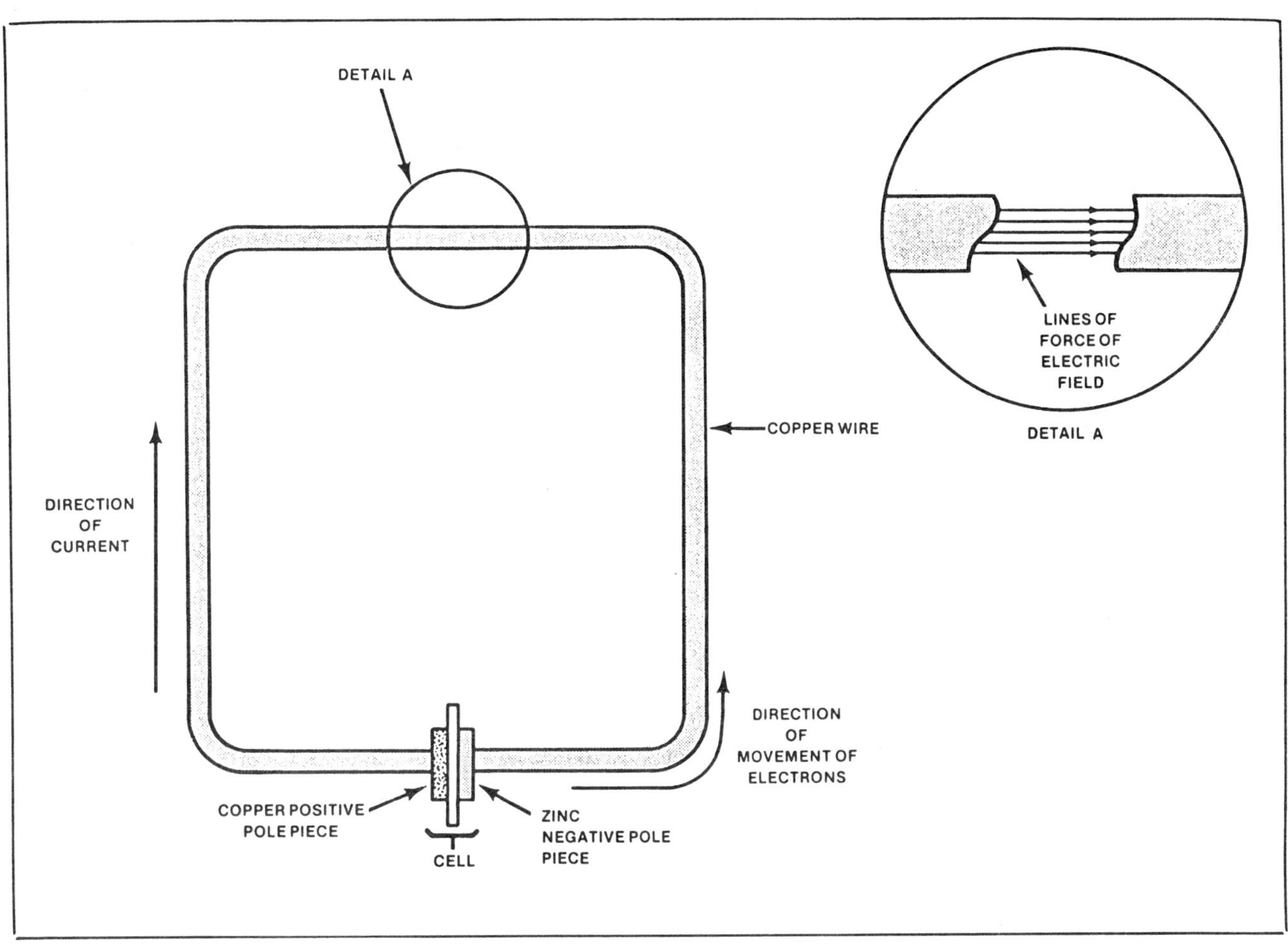

Fig. 2-12 The electric field within a copper wire connected to a voltaic cell.

due to the electric field, an electron leaves the positive pole to enter the liquid between the poles, and yet another electron leaves the liquid and is deposited on the negative pole, ready to start its round again under the influence of the electric field. Thus, because of the *electron pump* action of the cell, the potential difference between the positive and negative poles is maintained, even though we may require the electric field to do a moderate amount of work in propelling electrons through the wire.

In this particular case, the work done is the heating of the wire connecting the two poles of the cell. The potential difference between the two poles and the electric field will be maintained in the voltaic cell until such time as the electrochemical pump action comes to a stop, either through the formation of an insulating layer of gas bubbles on the positive electrode, or because the negative metal disk has been completely dissolved to furnish ions for the exchange process.

The ability of the voltaic cell to sustain its potential difference and electric field in spite of the fact that it is doing work, is one of the major differences between the effects of static and current electricity. A second important difference was discovered by the Danish physicist Oersted in 1820. During a university lecture demonstration of the voltaic battery, Oersted noticed the movement of a compass needle which happened to be near the wire connecting the positive and negative poles of the battery.

He was quick to conclude that the passage of electricity through a wire (an electric current as it came to be known) was associated with the creation of a magnetic field around the wire. This discovery provided a link between the study of electricity and magnetism which was not suspected earlier, and may be considered the start of modern research in *electromagnetic theory*.

Within a few weeks of Oersted's publication of his findings in a French scientific journal, the French physicist Andre Ampere discovered that the magnetic fields that exist around two parallel wires carrying electric currents interact with each other to produce a force between the wires. This force tends to push the wires apart if the electric currents are flowing in the *same* direction or, if the electric currents are flowing in *opposite* directions, the wires are forced together. In either case, Ampere discovered, this force depended upon the *distance* between the wires, and the *quantity* of charge that flowed past a point on the wires in a second. This latter quantity he called the *current.* The modern basic unit of current was named the *ampere* in his honor.

The discovery of electromagnetic effects, as current electricity is more correctly called, led to a temporary crisis of conflicting measurements. Sometimes it seemed as though it would be impossible to measure electrostatic and electromagnetic quantities with units that had any relationship to each other. Fortunately, it has been possible to devise a set of mks electromagnetic units, the so-called "practical system", which meets the everyday needs of most scientists, engineers, and technicians. The practical electromagnetic units are summarized and compared to the electrostatic units in Table 2-1, and will now be used almost exclusively in this book. The curious factor

$$\frac{1}{4 \pi e_0}$$

which appears in the practical unit formulas for potential and force, is necessary to make these units co-measurable with the electrostatic system. For the sake of computation,

$$\frac{1}{4 \pi e_0}$$

may be taken as 9×10^9. e_0, then, is equal to about 8.85×10^{-12}.

QUANTITY	Practical Electromagnetic Unit (emu)	Electrostatic Unit (esu)
Force between two charged bodies	$\vec{F} = \frac{1}{4 \pi e_0} \times \frac{q_a q_b}{d^2}$ where q_a and q_b are in coulombs, d in meters and F in newtons	$\vec{F} = \frac{q_a q_b}{d^2}$ where q_a and q_b are in statcoulombs, d in centimeters, and F in dynes
Potential of a 1 unit test charge a distance d from another charge q Potential difference	$V = \frac{1}{4 \pi e_0} \times \frac{q}{d}$ where q is in coulombs, d in meters, and V in volts 1 volt = $\frac{1}{300}$ statvolt	$V = \frac{q}{d}$ where q is in statcoulombs, d in centimeters, and V in statvolts 1 statvolt = 300 volts
Charge	Basic unit: the coulomb 1 coulomb = 3×10^9 statcoulombs (the charge carried by 6.2×10^{18} electrons)	Basic unit: the statcoulomb 1 statcoulomb = the charge carried by 2.08×10^9 electrons
Current	Basic unit: the ampere 1 ampere = a current of one coulomb per second (3×10^9 statcoulombs per second)	Basic unit: the statampere 1 statampere = a current of one statcoulomb per second

TABLE 2-1

A. Self-Test Questions:

1. Contact between a woolen cloth and a piece of amber, or a plastic comb, will transfer ___________ to the amber or comb, resulting in a ___________ charge on the amber or comb, and a ___________ charge on the cloth.

2. A charged body may be used to charge a neutral object some distance away from it by a process known as ___________.

3. Late nineteenth-century scientists though that the ___________ ___________ transmitted light and magnetic and electrical forces between two bodies not in contact.

4. Faraday's method of charting an electric field by means of lines of force is capable of showing the ___________ and ___________ of the field as well as the direction a small ___________ charge would move if placed at a point in the field.

5. A force is a ___________ quantity; to symbolize this we can draw a small ___________ above the letter representing the force.

6. Potential energy may be defined as a ___________ acting through a ___________. Mathematically this is symbolized as PE = ___________ × ___________.

7. The force on a test charge of one statcoulomb four centimeters away from another charge of one statcoulomb is ___________. (Don't forget the units.)

8. The potential at the point of the test charge in Question 7, assuming the test charge has been removed, is ___________ ___________.

9. A voltaic cell is composed of:

   a. ___________

   b. ___________

   c. ___________

10. One volt is equal to ___________ statvolt, one coulomb is equal to ___________ statcoulombs.

Briefly answer the following:

11. In what ways are gravity and electricity similar?

12. How does a lightning rod protect a house?

13. What is potential energy?

14. Describe the inverse square law. To what forces does it apply? To what quantities does it *not* apply?

15. What are the basic units of the mks and cgs systems? Name some other units, giving the system to which they belong.

16. Describe the function of a voltaic cell.

B. Questions For Thought:

1. Describe as many applications of electrostatic theory as you can. Are there certain areas of work that seem to be better suited to electrostatics than others? (Hint: look up the function of a van de Graff generator.)

2. What are the advantages of the metric mks system over the English (foot, pound, second) system? Consider the advantages and disadvantages of the change presently going on.

3. We have seen that stacking cells in a battery results in a higher potential difference between the endmost poles. What effect would increasing the diameter of the metal disks and leather or paper spacers have?

4. Given two charged bodies 40 cm (0.4 meter) apart with similar negative charges of 1 coulomb as shown in Fig. 2-13, calculate the potential at points A and B. What would be the strength and direction of the force on a positive test charge of one coulomb at point A? At point B? Notice that the potentials at points A and B from each charge may simply be added.

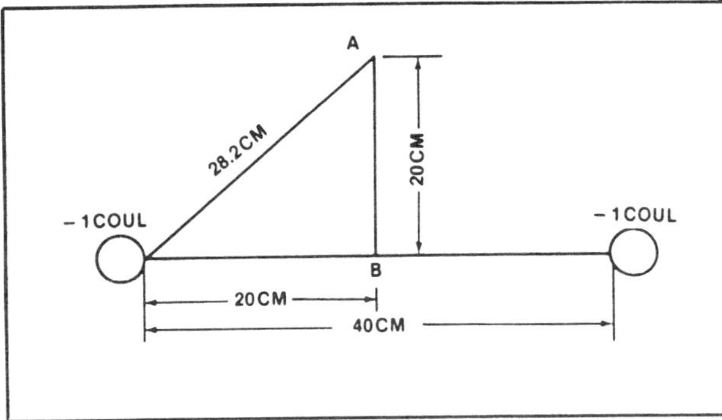

Fig. 2-13

C. Experiments and Demonstrations:

1. A fairly good indicator of the presence of charges or an electric field may be made by suspending a small styrofoam ball (1/4 inch in diameter is a good size) from a thread, or better still, a piece of very thin monofilament fishing line, as shown.

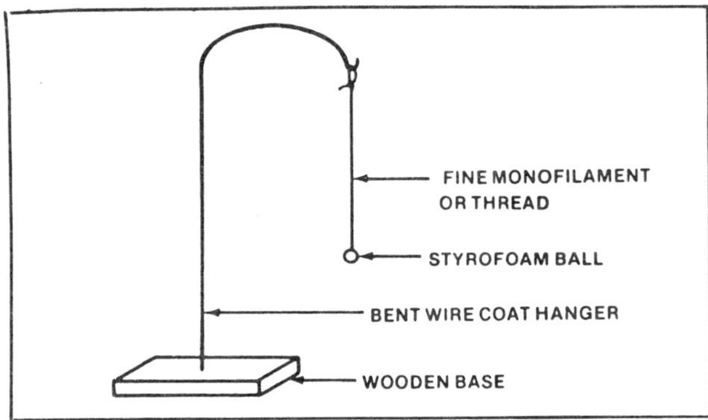

Fig. 2-14

a. Make two of these devices and test various combinations of materials, such as a rubber boot rubbed with a piece of fur, or a comb rubbed through your hair, for their ability to produce a static charge that can be seen in the deflection of the styrofoam ball when the boot or comb is brought near the ball.

b. Charge the two styrofoam balls by touching them with a comb that has been passed through your hair. Bring the balls toward one another and observe the results. Explain.

c. Charge one of the balls by contact, the other by induction (be sure to remove the grounding wire quickly, before removing the charged comb or rod). Bring the balls near one another and observe the results. Explain.

NOTE: The above experiments work best in dry weather.

2. A more sensitive indicator of electric charge, called an electroscope, may be made using a small jar, a bit of copper wire, a rubber stopper, and a bit of the aluminum foil stripped from the inner wrapper of a pack of cigarettes (Fig. 2-15).

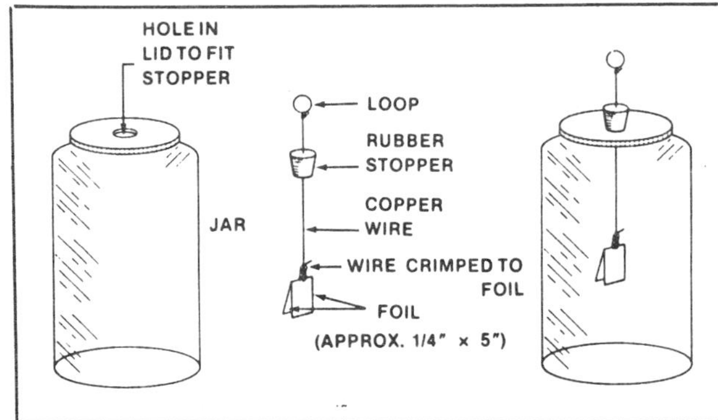

Fig. 2-15

a. When a charged object is touched to the loop at the top, the two foil leaves will repel each other, forming an inverted V. The width of the V will give a rough indication of the strength of the charge. The electroscope may be discharged by briefly touching the loop with a grounded wire.

b. Test the same materials used in experiment 1, noting the relative strength of the charges by observing the distance between the electroscope leaves. Be sure to discharge the electroscope before each test. (When the electroscope is discharged, the leaves should hang straight down.)

3. Construct a Leyden jar as shown in Fig. 2-10. Use aluminum foil in place of the silver foil and copper loops in place of a chain. Charge the jar by touching the ball with a wire attached to a static electric generator. Carefully bring a knuckle toward the ball. What length are the sparks that may be drawn from your model?

4. Using pennies, nickles, and disks of blotting paper soaked in a *dilute* solution of sulfuric acid, construct a voltaic battery of twenty or so cells (Fig. 2-11).

   a. With a voltmeter, measure the potential difference between the endmost coins. What is the potential difference of a single cell? Connect wires to the poles of a single cell and try to light a 1.5 volt flashlight lamp with it. Does the lamp light?

   b. Make another cell using a six-inch copper disk and a six-inch zinc disk. What is the potential difference between the two disks? Again try to light the 1.5 volt lamp. Will this larger cell light the lamp? (It should. Why?)

SECTION III

Schematics Cells In Series And Parallel

In Section 2 the ideas of potential difference and current were introduced in the discussion of a current electricity source such as the voltaic cell. In the course of his work, the avionics technician will come across many current electricity sources both electrochemical, such as the cell, and electromechanical, such as the generator. The basic principles discussed in these sections can be applied to all.

In this section we will begin to stress the practical side of our subject. While the earlier sections were necessary to develop a theoretical base, the following sections will feature more practical examples and exercises of the type the avionics technician will encounter in his daily work. Therefore, it is recommended that the reader attempt to answer the self-test questions and try all experiments. These self-test questions require a working knowledge of elementary algebra and exponentials, such as may be obtained from the mathematics textbook in this same Avionics Technicians Series. The experiments for this section require only the most common electronic components and a multimeter, which is a basic meter movement equipped to measure voltage, current, and resistance.

As a basis for the practical material to follow, this section begins with a discussion of DC and AC sources followed by an introduction to schematic and wiring diagrams. These are the conventional ways of describing how sources of current electricity and the components to which they are connected may be arranged. Also defined are open and closed circuits. The section concludes with a discussion of how cells may be connected to form batteries like the familiar 12-volt storage batteries used in automobiles. In this discussion, the concepts of series and parallel, aiding and bucking, and resistance are introduced.

A. DC and AC Sources

The direct current (DC) source, whether it is a cell of the sort discussed in Section 2, a group of cells connected to form a battery, or an electromechanical device, has two output poles, one of which is always positive, the other always negative. In general, the potential difference across the source is made to remain as nearly constant as possible.

The alternating current (AC) source, which will be considered in greater detail in a later book in this series, is an electromechanical device with two output poles, each of which regularly changes from being the positive pole to being the negative pole. When a standard AC source device is started, there is no potential difference between the two poles. As time elapses, the potential difference between the poles increases to a maximum, with one of the output terminals serving as the positive pole, and the other as the negative. After maximum potential difference is reached, it begins to decrease, finally reaching zero, (i.e., no potential difference between the poles). At this point, the polarity of the output terminals reverses and what was the negative pole becomes the positive pole, with the former positive pole becoming the negative pole. The potential difference again builds up to a maximum (this time with the electric field direction reversed) and then falls back to zero. Then the polarity of the output terminals again changes, returning to its initial state. This whole process of events is called one complete *cycle* of an AC source. The entire cycle repeats itself 60 times per second in American standard house current and 400 times per second in most airborne AC supplies.

In this book, we shall deal with the electrical effects produced by connecting various electronic parts, called components, to terminals or poles of a dc source. For the most part we will ignore the effects that might be noted at the moment the components are connected. Instead, we will concentrate on those effects each system shows after "settling down" to what are called steady state DC conditions. This does not mean that the switch-on or so-called transient state effects are to be ignored. Rather, it will be easier to discuss them along with the electrical effects observed in components connected to AC sources in another volume of this series.

B. Pictorial, Wiring, and Schematic Diagrams

At the beginning of our exploration of avionics equipment, it is best to agree on the type of maps we will be using and how we will interpret them. These maps are indeed necessary, for there is no way that the winding conduction paths and masses of components in even the simplest electronic equipment can be usefully described using words alone. In fact, we might modify the old saying to: "a picture may be worth a thousand words, but a map is beyond price."

Just as there are specialized maps for pilots, motorists, weathermen, and politicians, there are a number of different kinds of diagrams used in electronics, depending upon the requirements of the user. The simplest to read is called the *pictorial diagram.* It may be either a line drawing (sometimes with shading to emphasize shapes), or even a photograph of a piece of equipment. The names, part numbers, or what are called *reference designators* are shown, either printed on the components or connected to them by thin lines called *leader lines.*

Reference designators, which generally consist of a combination of letters and numbers, are used to uniquely identify the component or assembly to which they refer. For this reason, reference designators are "stackable", that is, a full reference designator identifies the major assembly, sub-assemblies, and, finally, the component. Let us assume that a particular piece of equipment, a cabinet housing a transmitter in a control tower, contains two power supply units, and that each power supply unit contains three identical sub-assemblies. In this case, the reference designator of the entire cabinet could be HFT 1 (high frequency transmitter number one) and the two power supplies could be called PS1 and PS2. The full reference designators of the two power supplies in the cabinet would then be HFT1PS1 and HFT1PS2. Similarly, if the three subassemblies are called A1, A2, and A3, the reference designator HFT1PS2A3 uniquely refers to the third subassembly of the second power supply of high frequency transmitter number *1.* The method may seem cumbersome at first, and even a little silly, but it is much simpler and much less confusing to tell a technician to "interchange HFT1PS1A3 and HFT1PS2A3" than it would be to say the same thing without reference designators.

In general, then, a pictorial diagram provides us with information concerning the overall appearance of a unit, the shapes, relative sizes and location of components, interconnecting wires and cables, and so forth. It usually names the components in such a way as to permit cross checking with a parts list. However, it does not provide any clue to the function of any part within the unit or system. An additional form of pictorial is sometimes seen, the assembly or parts blow-up. It is a curious sort of diagram, showing each individual component or mechanical part of the unit connected to its eventual place on a main assembly or chassis by a dotted line. The overall appearance of this drawing is that of a unit that has had an explosive charge go off inside it. In spite of its bizarre appearance, this sub-type of pictorial diagram is of particular value for showing otherwise hidden components or assemblies. It may, if carefully done, greatly help in the disassembly or assembly of the equipment. See Fig. 3-1 for examples of pictorial diagrams.

A second kind of map or diagram used in electronics is the *interconnect* or *wiring diagram.* This diagram is meant especially for those who must wire or interconnect equipment. In general, the wiring diagram uses squares or rectangles to represent the various components or units involved, and is detailed only in specifying the wires and connection points. As might be expected, the colors of the insulation on the various wires, wire sizes, and cabling details are given.

For simple units, the pictorial diagram is often made to serve as a wiring diagram (as shown in Fig. 3-2). Again, as in the case of the pictorial, no effort is made to show the function of the components. The main purpose of the wiring diagram is to describe the electrical interconnections of the various components that make up a piece of equipment.

The third kind of map with which we shall deal in this book is a functional one, the *schematic*

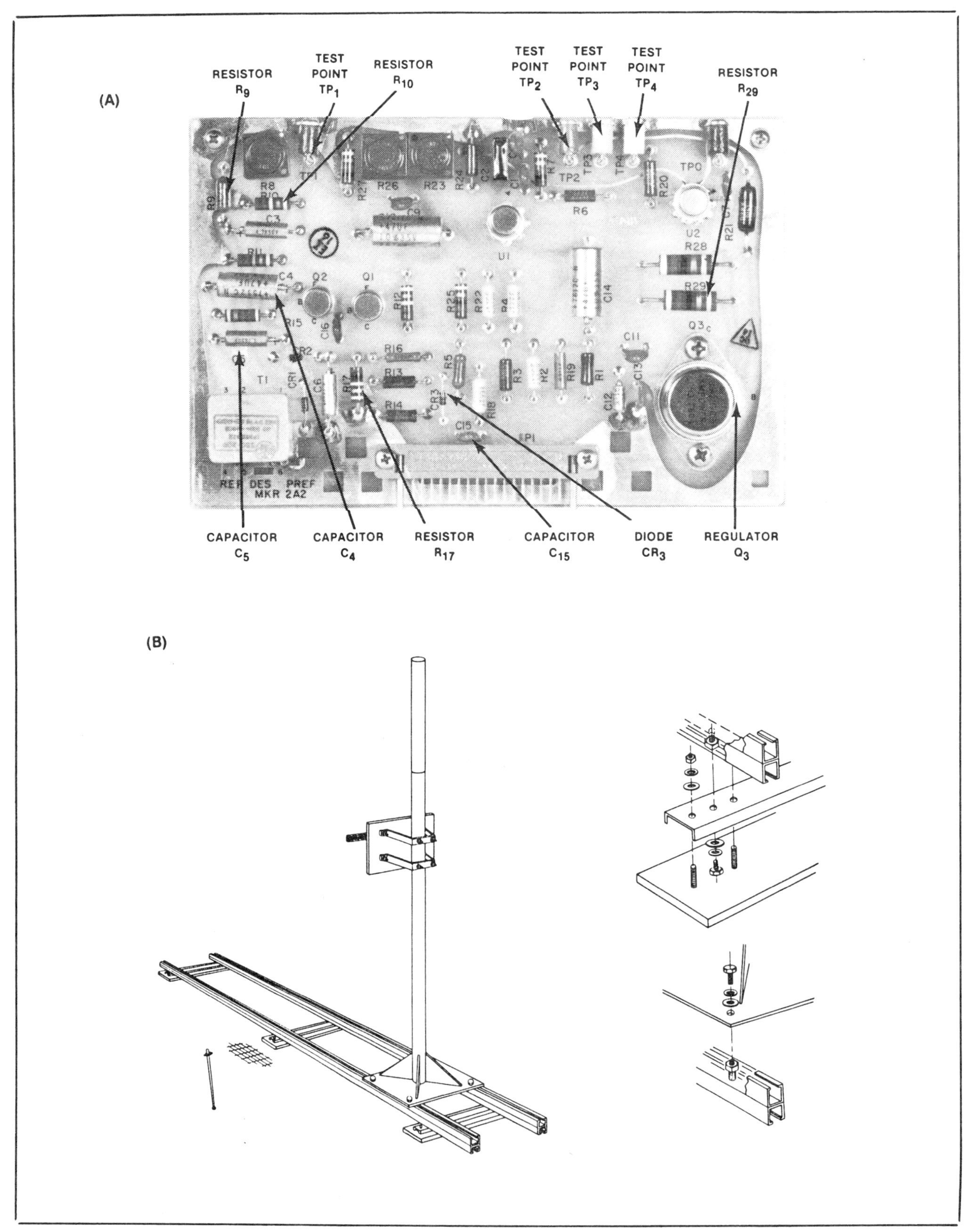

Courtesy of Wilcox Electric

Fig. 3-1 A — Pictorial diagram of a printed circuit board showing some components. B — Assembly drawing and parts blow-up.

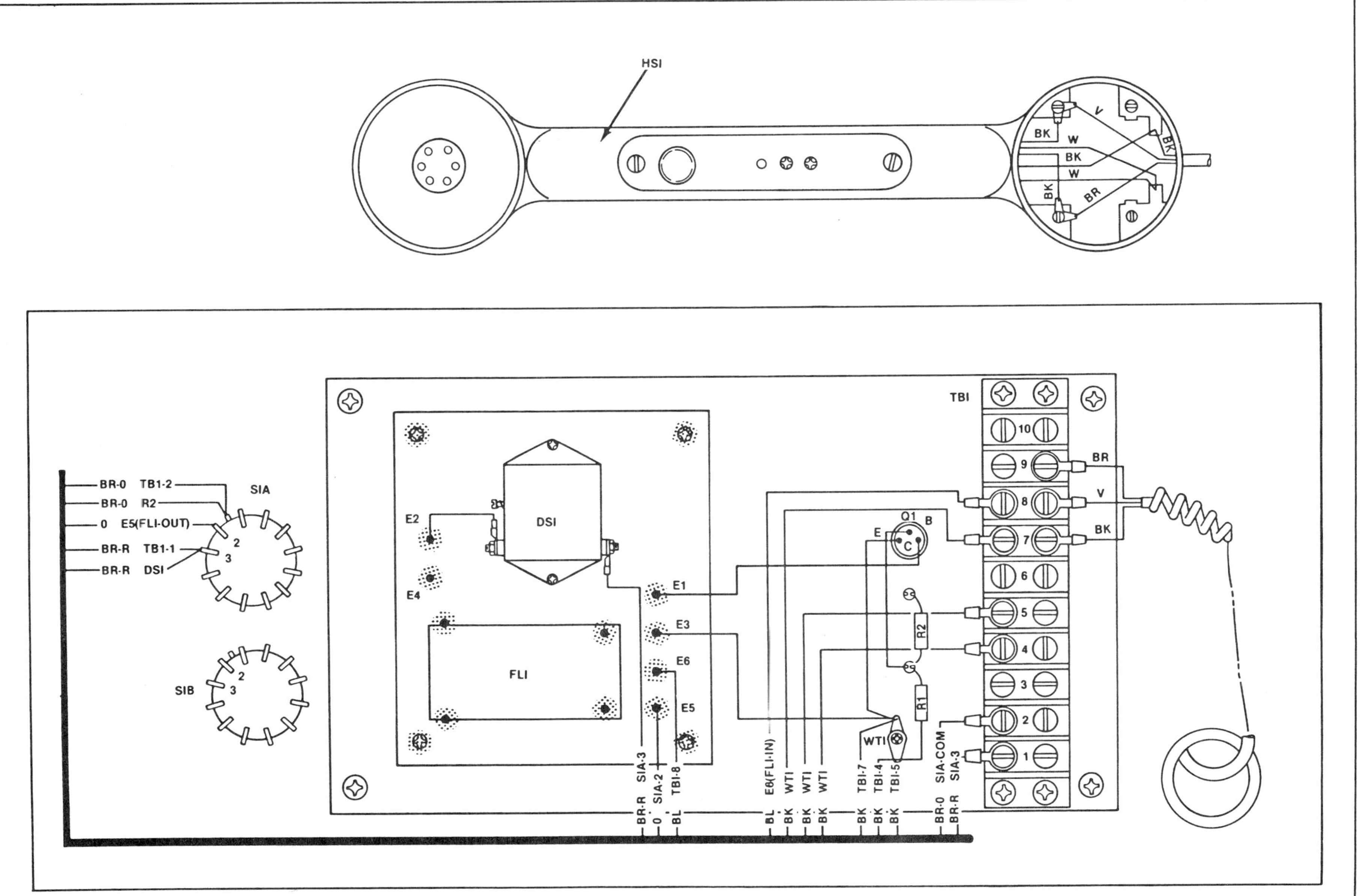

Courtesy of Wilcox Electric

Fig. 3-2 Pictorial wiring diagram.

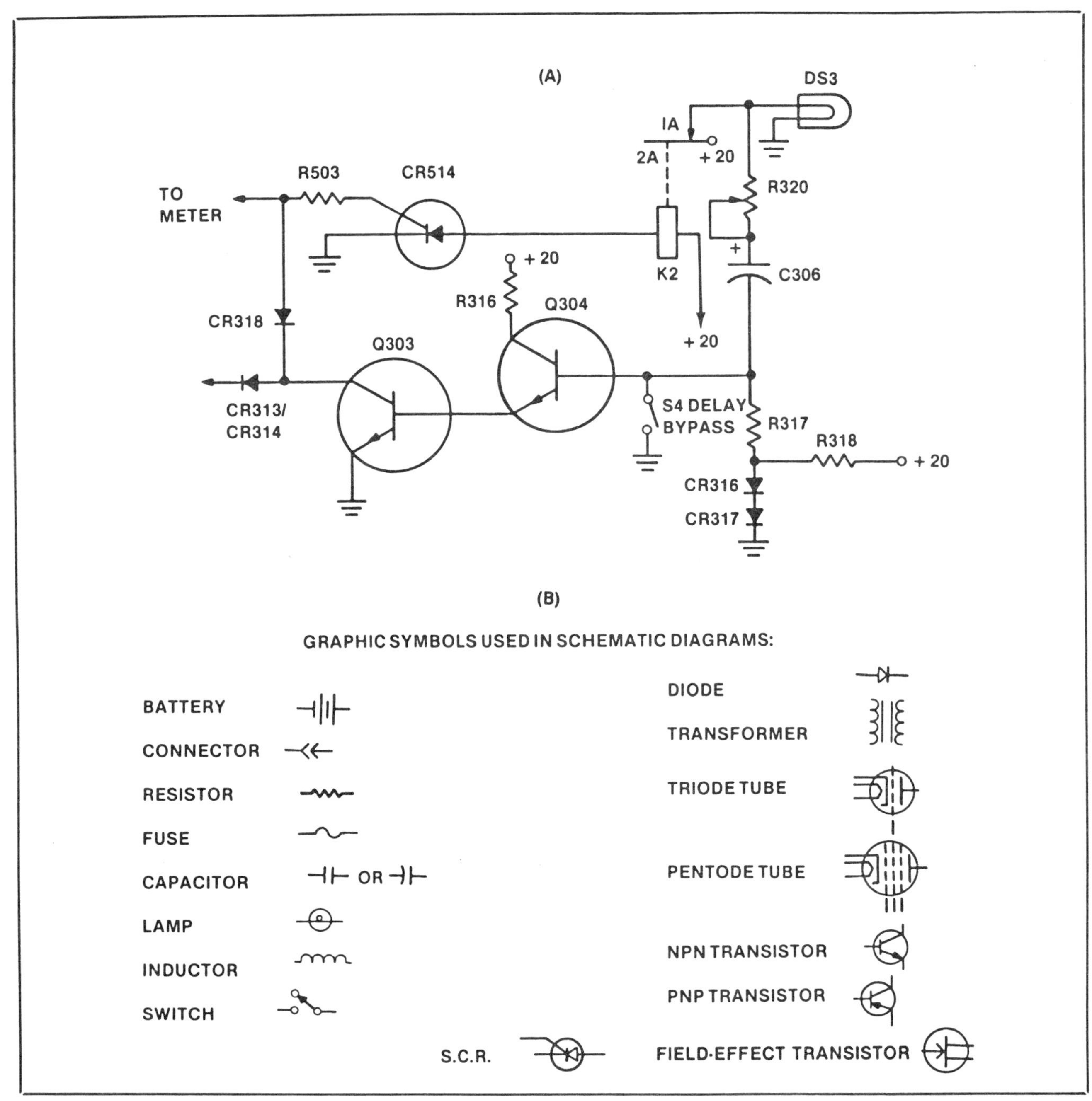

Courtesy of Wilcox Electric

Fig. 3-3 A — A sample schematic diagram showing a small portion of the entire unit. B — Graphic symbols used in schematic diagrams.

diagram. The schematic diagram is an arrangement of standardized graphic symbols for the functions of the parts of a piece of equipment. The symbols are connected by solid lines representing the electric current or potential paths between components (these lines sometimes represent actual wires, but sometimes they do not; as long as an electrically conductive path exists between two points on a schematic diagram, a line is drawn between them).

Once we become familiar with the conventional symbols and layout of a schematic diagram (Fig. 3-3), a surprising amount of information about the function of the unit becomes immediately available. The basic graphic symbols for various components are given in Fig. 3-3. These will be repeated and new symbols introduced as needed.

In general, the reference designators and functional quality magnitudes (the resistance of resistors, the voltage of batteries, etc.) are provided for each component in the schematic diagram. But no indication is given of the physical size or shape of the units. This is especially worth remembering because the schematic will group the various components into functional sub-units, regardless of their actual physical location in the equipment.

When a schematic diagram becomes so complex that it is hard to follow the major current flow paths, a *simplified schematic,* leaving out all but the most important components and paths, is used. Very often the diagrams that serve as illustrative examples in engineering textbooks are of this kind. Many an unwary student, the author included, has been fooled into building the equipment described by such a schematic, only to find that it would not work!

Another sort of simplified schematic, called the block diagram, is sometimes used for explaining complex equipment. The block diagram substitutes labelled boxes for functional sub-units and shows the generalized paths of control and other signals between the boxes. Fig. 3-4 shows the block diagram of a portion of an instrument landing system.

There is at least one more important type of map or diagram which the avionics technician must be able to use — the logic diagram. However, we will not meet with such diagrams in the earlier books in this series.

C. *Open and Closed Circuits*

Having pointed out the kinds of maps that may be drawn to illustrate part interconnections, we may now use these maps to discuss the interconnections themselves. It is useful to consider the interconnections between parts of an electronic device as the means of channeling the energy, that is, the potential difference or the electric current, to a place where it will perform some function.

Although there are important exceptions, especially in the field of microwaves, electrical energy is channeled from place to place by metallic, usually copper, conduction paths which may be wires, or strips of metal printed on an in-

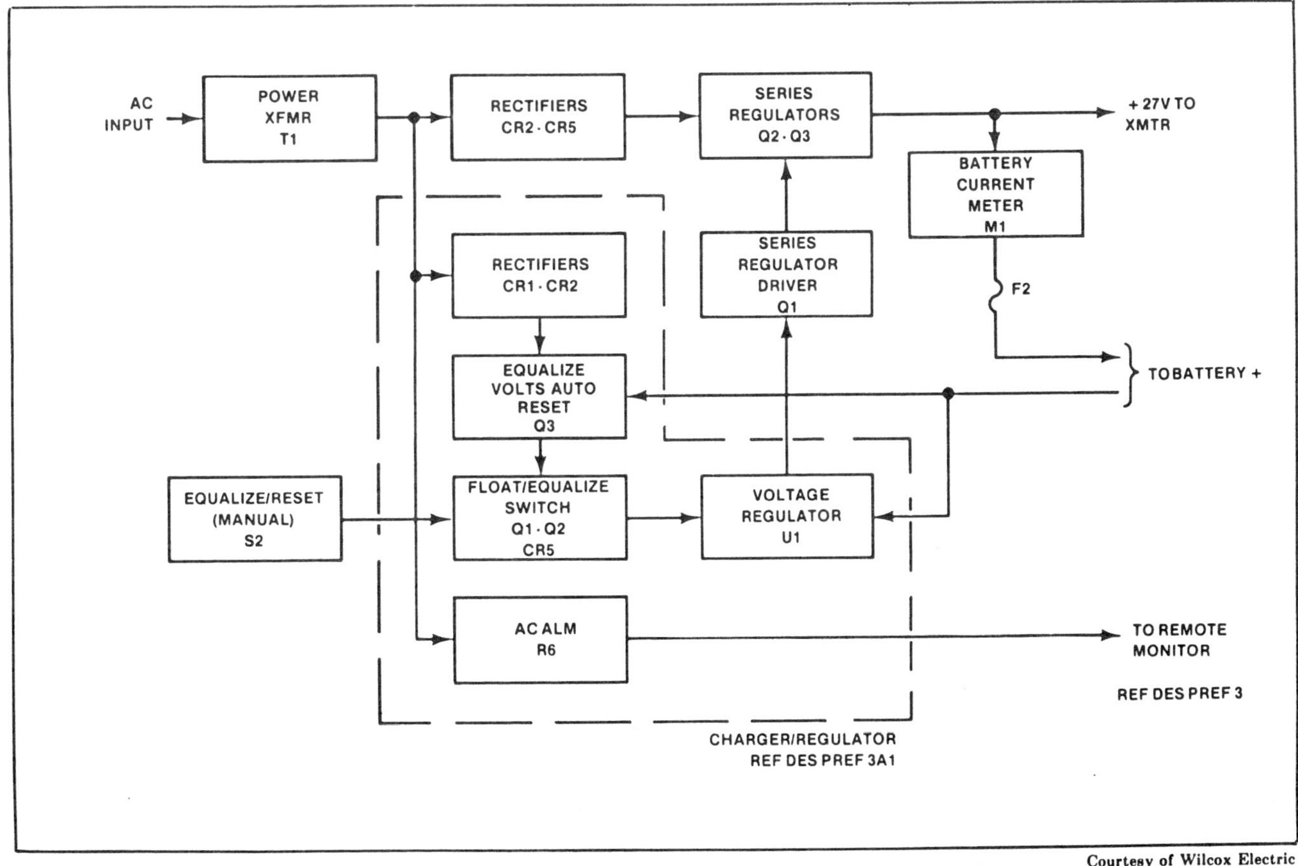

Courtesy of Wilcox Electric

Fig. 3-4 Block diagram of a transmitter power supply.

sulating support. We use the word *circuit* to refer to an entire complex of conducting paths and components. The term "circuit" is very general and may refer to anything from a simple piece of wire connected between the two poles of a cell, to a complex branching and looping path involving dozens of components and many wires.

In Section 2 it was shown that a cell is both a source of voltage and an electron pump. Therefore, the sort of circuit we connect to a cell depends on which of the two characteristics of the cell we wish to utilize. In the past, it was stressed that any circuit capable of performing a useful function had to possess a continuous conductive path to allow the electrons to flow from one pole of an energy source to the other. This is something of an oversimplification.

It is true that if we wish to perform some sort of work with a flow of electrons, we must provide what is called a *closed circuit*, or a continuous path from the negative pole of a cell or other source, where there is a surplus of electrons, to the positive pole, where there is a deficiency. Once this path is established, the electric field (caused by the potential difference or voltage across the poles of the cell) will cause electrons to begin moving around the closed circuit. Breaking this closed circuit, by actually cutting the conduction path, will stop the flow of electrons. A circuit which does not provide a continuous conduction path for electrons is called an *open circuit* (Fig. 3-5).

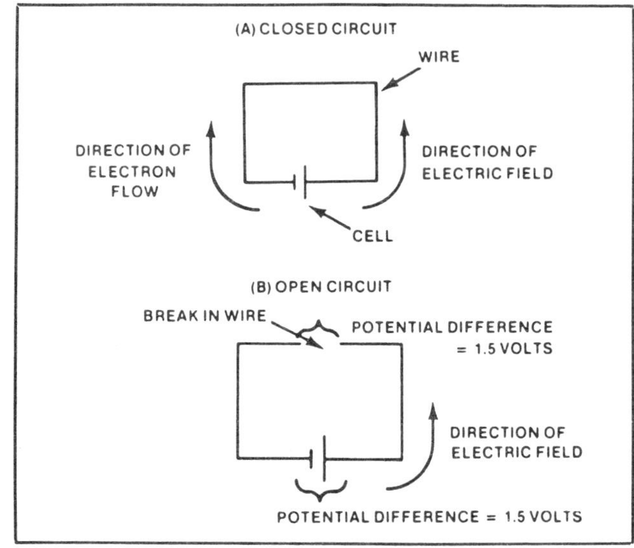

Fig. 3-5 Illustration of the differences between closed and open circuits.

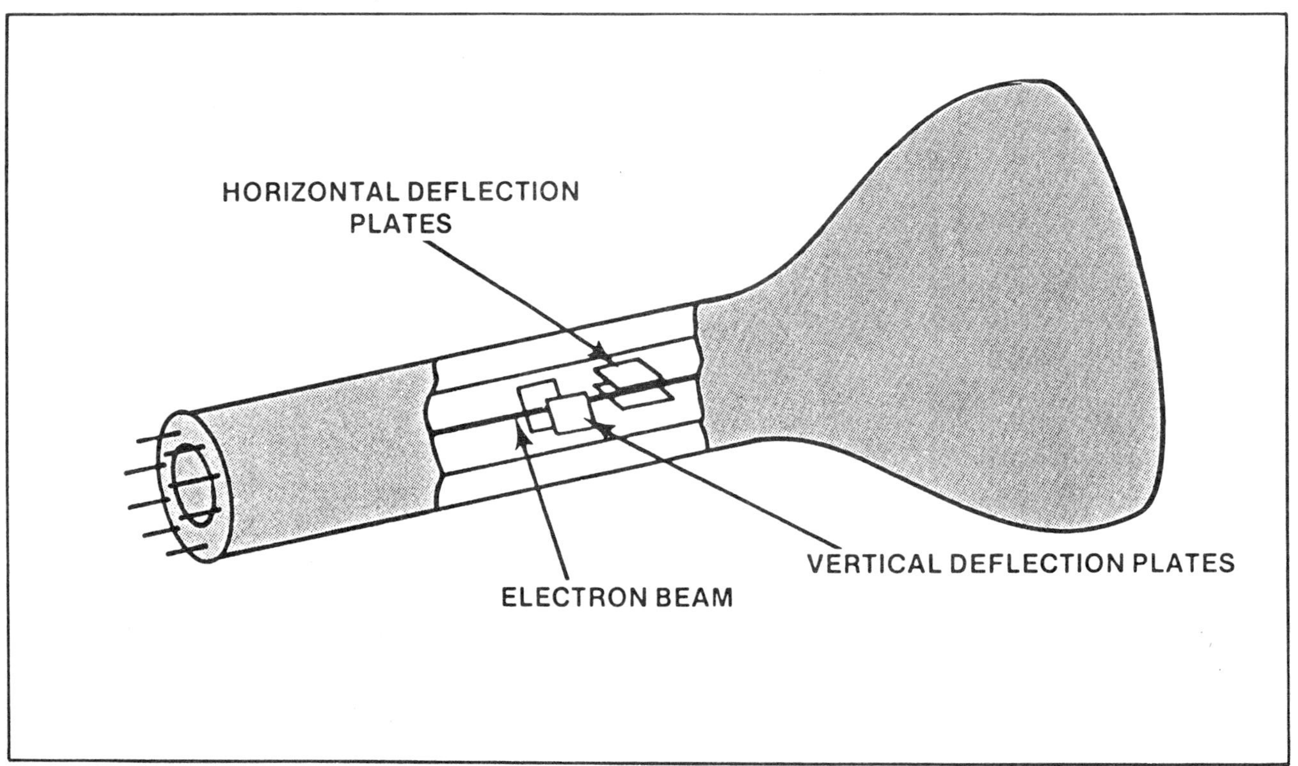

Fig. 3-6 Oscilloscope tube deflection plates. Although there is an open circuit between each pair of plates, a potential difference across them produces a useful effect.

Although there is no current flow in an open circuit, it is still capable of performing many useful functions when a potential difference alone is required. In the cathode ray tube (Fig. 3-6), the familiar picture tube of the oscilloscope, two pairs of metal plates, each attached to one pole of a variable power supply (the input of the oscilloscope) are made to deflect a beam of passing electrons. In this case, the electric field caused by the potential difference between the plates exerts a control influence, neither a current flow nor a closed circuit conduction path in the traditional sense is necessary.

A number of important differences exist between the use of the potential difference and the electron pump aspects of a current electricity source. A current produced by a DC source can be completely stopped by even the smallest nonconducting break in the conduction path between the positive and negative poles. Although there will be no current flow, a potential difference will exist across this break in the conduction path.

To understand this, imagine that both poles of a cell are to be touched by the ends of a long heavy wire (Fig. 3-7A). Before the ends of the wire are touched to the cell, there is a very weak electric field in the wire caused by the weak electric field surrounding the cell. Although this field tends to follow the path of the conducting wire, it is too weak to move a significant number of electrons in the wire itself, and there is no potential difference between the ends of wire. When the ends of the wire are touched to the cell, an electric field which is completely contained within the wire begins to pass along its length with the speed of light (3×10^8 m/sec). The conduction electrons in the wire begin to move because of the force exerted on them by the field, but the speed of movement of an electron is much less, about 1.5×10^{-4} m/sec.

The electron, although tiny, is a physical entity. It is therefore subject to collisions with the atoms in the wire. In a conductor, like a copper wire, all the conduction electrons are in constant movement in all directions. What the electric field does is modify this random movement so as to cause all the electrons in the wire to undergo a net drift at 1.5×10^{-4} m/sec in the direction opposite to that normally accepted as the direction of the field. (Remember, the direction of the field is the direction that a *positive* charge would move, if placed in the field.)

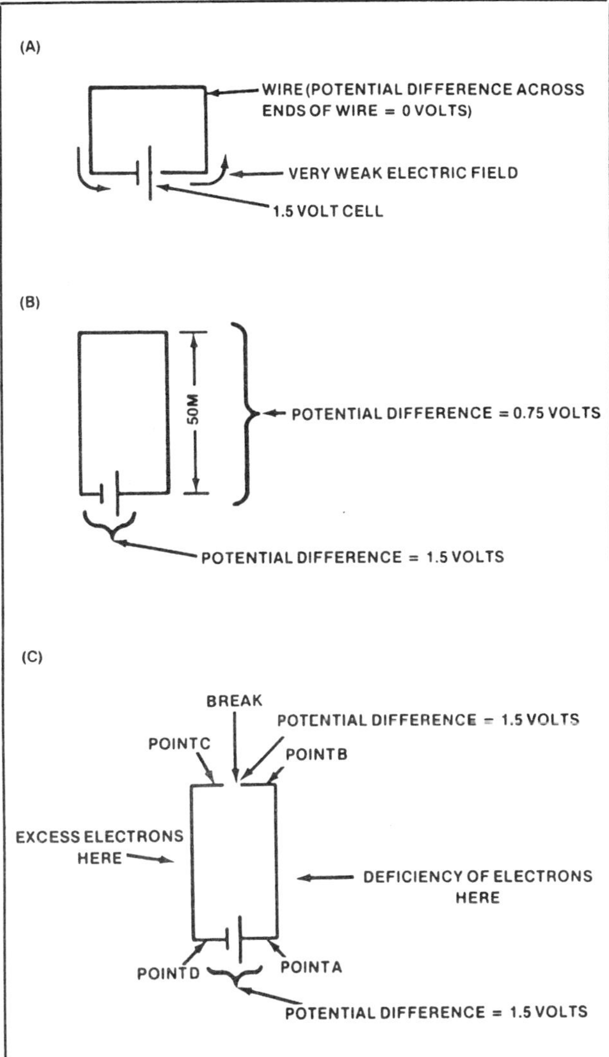

Fig. 3-7 Potential difference and electron flow current.

If the wire in Fig. 3-7B is 100 meters long, and the potential difference across its ends is 1.5 volts, what will be the potential between either end and a point 50 meters away or halfway along its length? If your answer is 0.75 volts, you would be correct, for the potential difference between any two points along the wire depends upon where the two points lie on the potential gradient that exists between the two ends of the wire. If however the wire is cut (Fig. 3-7C), the potential difference between points A and B and between A and C will *not* be 0.75 volts. After a very brief time all the available conduction electrons in the segment of wire attached to the positive pole will enter that pole. At this time, the potential difference between points A and B will be zero. As electrons enter the positive pole of the cell, a similar number of electrons will enter the segment of wire connected to the negative pole, and will "pile up" there, until the potential difference between points B and C (which will be the same as the potential difference between points A and D) will be 1.5 volts. In this example, we see that an open circuit may be considered as a way of moving a potential difference from a source to a point where it may be utilized for control purposes.

In other words, a closed circuit provides a path for a flow of electrons, that is, a current, while an open circuit provides a path for a potential difference. Since there is no movement of charges in an open circuit once a steady state has been reached, no further work is done by the cell.

D. Cells, Batteries, and Voltage

The voltaic cell mentioned in Section 2 represents only one of the possible sources of potential difference. Its modern version is unrivalled as a compact, portable source of energy. It exists in many forms — including rechargeable nickel-cadmium cells which do not dissolve their negative pole piece as do voltaic cells or the familiar *dry cells* widely used in flashlights or transistor radios. Regardless of its forms, make-up, or size, the schematic symbol for a single cell is —|⊦± .

As often as not, the tiny plus sign marking the longer line as the positive pole is omitted. This polarity is considered standard. The long line is always the positive pole.

As stated previously, the potential difference between the positive and negative poles of the cells currently in use varies somewhat depending on the type. However, it is generally in the neighborhood of 1.5 volts. If the positive pole of one cell is connected to the negative pole of another, as in Fig. 3-8, the potential across the two will be equal to the sum of the potentials of the two cells separately. We call this a *series aiding* connection of the two cells and this is the way in which larger potential differences than those offered by a single cell may be obtained (as in the voltaic pile). The familiar 4.5 volt electric lantern battery is composed of three 1.5 volt cells connected in such a manner and packaged together by the manufacturer.

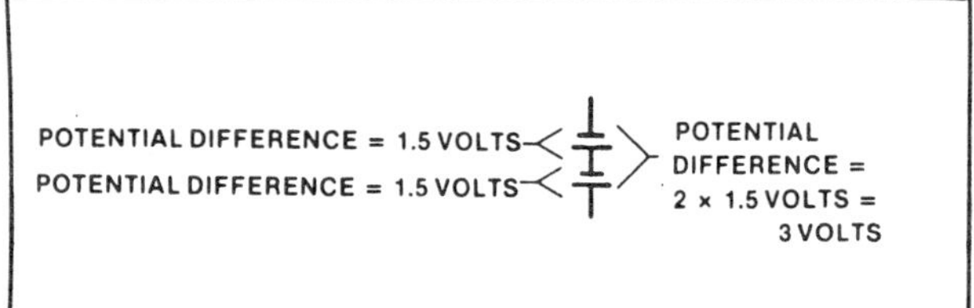

Fig. 3-8 Two chemical cells connected in series-aiding yield a higher potential difference.

But what happens if the two cells are connected, as in Fig. 3-9A? The difference between this arrangement and that of Fig. 3-8 becomes clear if we connect a conducting wire across the combination as in Fig. 3-9B.

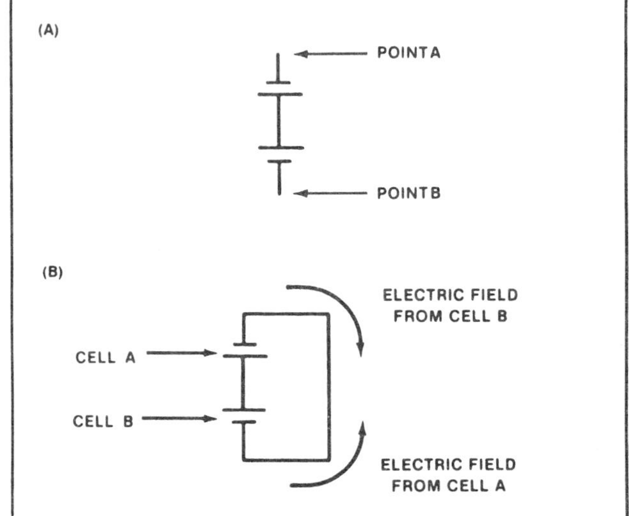

Fig. 3-9 The series bucking connection.

Since the electric field caused by a potential difference is considered as directed from the positive pole to the negative pole (i.e., the direction opposite to the flow of electrons), we can see from the diagram that the two fields will cancel each other. If the potential difference across each of the cells in Fig. 3-9A is the same, there will be no potential difference between points A and B, and no current will flow through the wire. If the potential difference across one cell is greater than the other, there will be a net potential difference which is the difference between the voltage measured across cell A and across cell B. The direction of current flow will be in the direction determined by the cell with the greater potential difference. This arrangement is called *series bucking*.

The schematic symbol for cells connected in the series aiding configuration is —|ı|ı⊢ . The number of cells usually drawn does not, however, really picture the number of cells so connected. As mentioned earlier, this arrangement is called a battery.

E. *Cells, Batteries, and Current*

If a battery formed of four small, penlight cells connected in series aiding (or in series as this arrangement is more usually called) is connected to a six-volt automobile headlight lamp, little or nothing will appear to happen. True, if we had measured the voltage across our battery before connecting the lamp, the meter would have registered six volts. Why, then doesn't the lamp light up, instead of giving out, at best, a feeble red glow?

The answer to this lies in the type of work we are asking our battery to perform. Basically, the electric lamp is a device which makes use of a property of conductive materials which we have not yet introduced. This property is *resistance.* All conductors, to a greater or lesser extent, possess this property of resisting electron flow through them. The resistance of the conducting path will, in a closed circuit, turn the energy supplied by the cell into heat which is caused by the passage of the current of electrons (discussed at length in Section 4). If a thin wire of a material with a high resistance is connected between the poles of a cell, it will get hot and begin to glow. If the flow of electrons is sufficient, the wire will get white-hot and be capable of giving off a significant amount of light. This is exactly what happens in an electric lamp.

The small battery in our example simply isn't able to furnish enough current to heat the relatively heavy wire (called the filament) of an automobile headlight sufficiently.

If four penlight cells (Fig 3-10) are connected in such a way that all the positive poles are connected to one wire and all the negative poles are connected to a second wire, the current that can be obtained from this *parallel* connection will be four times greater than the current that can be obtained from a single penlight cell. This allows each cell to contribute its share of the current. The voltage across points A and B will be 1.5 volts, the same as the voltage across a single cell. This is called a *parallel* arrangement.

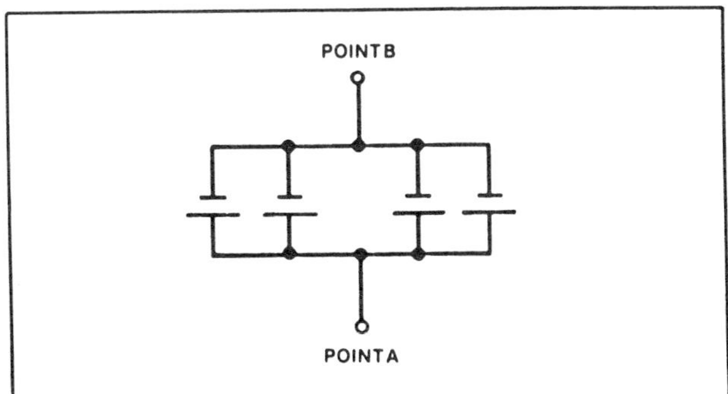

Fig. 3-10 Four cells in parallel. This connection results in a higher current than could be obtained from a single cell.

In the parallel arrangement, although the current supplied by the parallel cells increases, the voltage across the battery remains the same as that across a single cell. This is exactly the *opposite* of the series arrangement, where the voltage across the battery is increased, while the current supplied remains the same as the current supplied by a single cell.

F. *Series-parallel and Parallel-series Batteries*

Using a combination of the series and parallel cell arrangements, it is possible to construct a battery that will furnish more voltage *and* more current than a single cell. Given eight cells it is possible to construct two four-cell, parallel arrangements which can be connected in series (Fig. 3-11). This is called a *series-parallel* arrangement, since it places two parallel batteries in series.

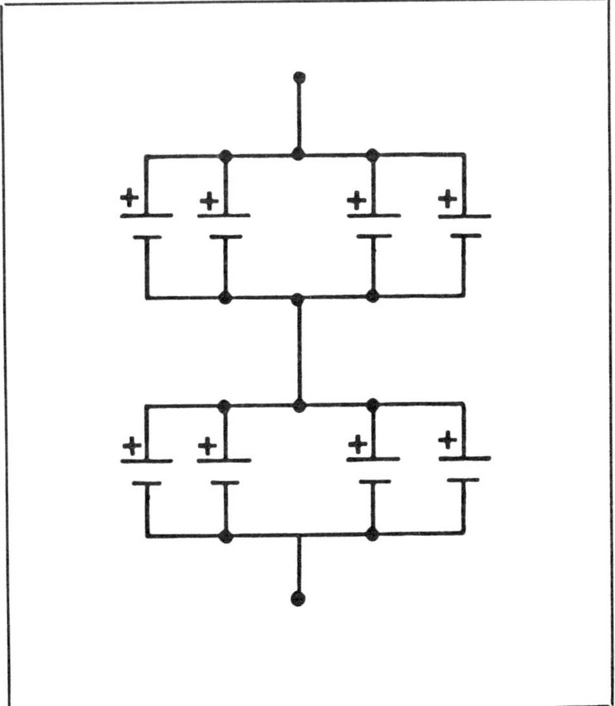

Fig. 3-11 A series-parallel circuit — the series connection of two parallel groups.

Similarly, we may construct four series batteries of two cells each, and then put all four in parallel (Fig. 3-12). This is called a *parallel-series* arrangement. If each cell is capable of delivering a maximum of 0.5 amperes (commonly abbreviated as *amp or a*), and has a potential difference of 1.5 volts across it, the total voltage across both batteries will be 3 volts and the maximum current capability 2 amps, disregarding losses due to the resistance of the cells themselves.

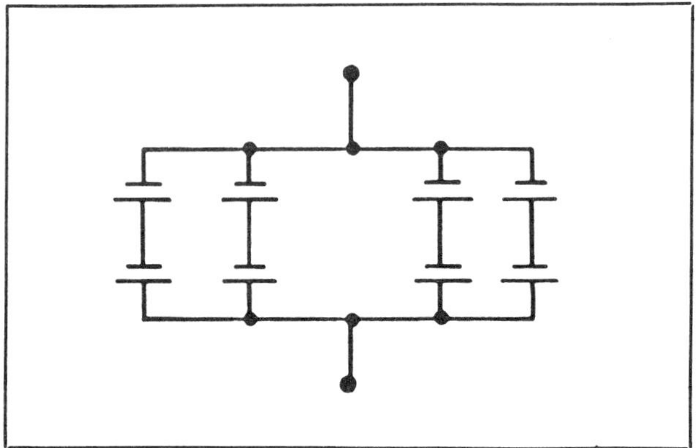

Fig. 3-12 A parallel-series circuit formed by paralleling four series strings.

A slightly different arrangement, pictured in Fig. 3-13, is capable of furnishing a potential difference of 6 volts, and has a current capability of 1 amp. In every case, it should be noted, each cell is capable of pumping 0.5 ampere and each furnishes a potential difference of 1.5 volts.

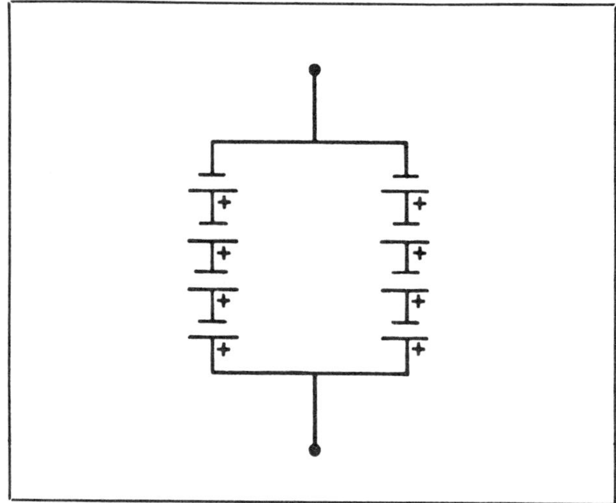

Fig. 3-13 A parallel-series battery. If the individual cells each furnish a potential difference of 1.5 volts and a current of 500 mA, the battery will furnish 6 volts and a current of 1A.

The maximum *power* that can be furnished by a cell or a battery is measured in *watts*. This value is equal to the product of the maximum current capability times the voltage across the cell or battery. The maximum power furnished by the batteries of Figs. 3-11 and 3-12 is then 3 volts × 2 amperes, or 6 watts. The maximum power furnished by the battery of Fig. 3-13 is 6 volts × 1 ampere, or 6 watts.

From these examples it is possible to see that although we can juggle the voltage-current relationship of a battery by arranging the cells differently, the maximum total power delivered by a number of cells can be no greater than the number of cells times the power available from a single cell.

A. Self-Test Questions:

1. A DC source has ________ output terminals, one of which is always the ________ terminal, the other is always the ________ terminal.

2. A ________ diagram shows the relative sizes, shapes, and locations of components in a unit.

3. A ________ ________ is a combination of letters and numbers which is used to uniquely identify an assembly or a component.

4. The ________ diagram conveys the most information about the function of the various components; it is made-up of graphic ______.

5. The symbol ⊣|ı|⊢ represents a ________.

6. A ________ circuit may be used to transfer a potential difference from a battery to a place where it is required.

7. A ________ circuit permits the flow of ________________.

8. A current does not exist in a ____________ circuit.

9. More current than is available from a single cell is obtained in a ______________ arrangement; an increase in potential difference is obtained from a ______________ circuit.

10. Two equal cells arranged in a ________ ________ arrangement produce a potential difference of zero across the battery thus formed.

B. Questions for Thought:

1. Which of the types of diagrams discussed in this section would you absolutely require if you were:

   A. an assembler building a unit?

   B. an engineer working out an idea for a new kind of aircraft anti-collison system?

   C. a technician installing a new airborne radio in place of an older model?

   D. a technician repairing an inoperative unit?

   What would you expect to obtain from the diagrams you have chosen in each case?

C. Experiments and Demonstrations

The following experiments are designed to familiarize the reader with the various ways in which cells may be arranged, and with the various diagrams used to describe circuits. The materials required for these experiments are:

1) eight penlight cells (size AA), BT_1 through BT_8;

2) eight single battery clips (small plastic clips designed to hold single AA cells) or equivalent for facilitating connections to the cells;

3) a multimeter with a current scale permitting measurement of 5 amps;

4) various lengths of wire;

5) four screw-type or spring clip terminals (E_1 through E_4).

1. Figure 3-14 gives a wiring list and pictorial diagram for a series battery.

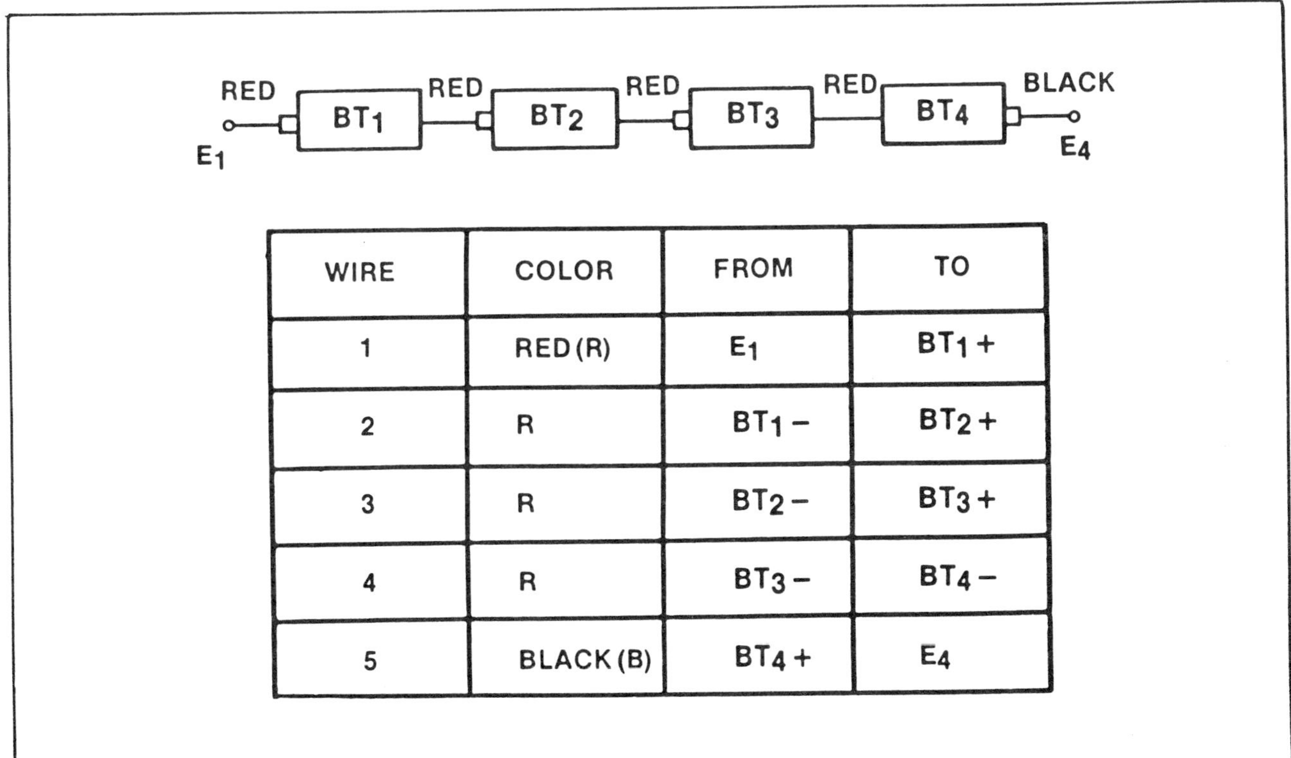

WIRE	COLOR	FROM	TO
1	RED (R)	E_1	BT_1+
2	R	BT_1-	BT_2+
3	R	BT_2-	BT_3+
4	R	BT_3-	BT_4-
5	BLACK (B)	BT_4+	E_4

Fig. 3-14

a. Draw the schematic.

b. Calculate and then measure the potential difference (voltage) across terminals E_1 and E_4.

2. Construct the battery described by the following wiring table:

Wire	Color	From	To
1	red	E_1	E_2
2	red	E_2	BT_1+
3	white	E_2	BT_3+
4	white	BT_3-	BT_4+
5	white	BT_4-	E_3
6	red	BT_1-	BT_2+
7	red	BT_2-	E_3
8	black	E_3	E_4

a. Draw a wiring diagram and a schematic diagram of the circuit.

b. Calculate the voltage between E_1 and E_4. Measure the voltage using the multimeter.

c. Using the multimeter on a high current scale, briefly touch the positive (red) test probe to E_1 and the negative (black) test probe to E_4. The current measurement is called the short circuit current capacity of the battery you have constructed. Be careful to make this measurement quickly, since a continued short circuit will seriously damage the cells.

3. Construct at least five different batteries using up to eight penlight cells, battery clips, terminals and wires. Draw the schematic diagram of each arrangement. Measure terminal voltage and short circuit current of each battery.

a. Which of the batteries you have constructed are equivalent? (HINT: consider the voltage and current capacity of each arrangement.)

b. Which of the batteries you have constructed produce the greatest power (current times voltage) per cell?

Caution: Keep all short-circuit current measurements as brief as possible using the highest practical multimeter current scale (10 amps full scale recommended).

4. Using the chart given in Fig. 3-3, identify all the components in the schematic of Fig. 3-15.

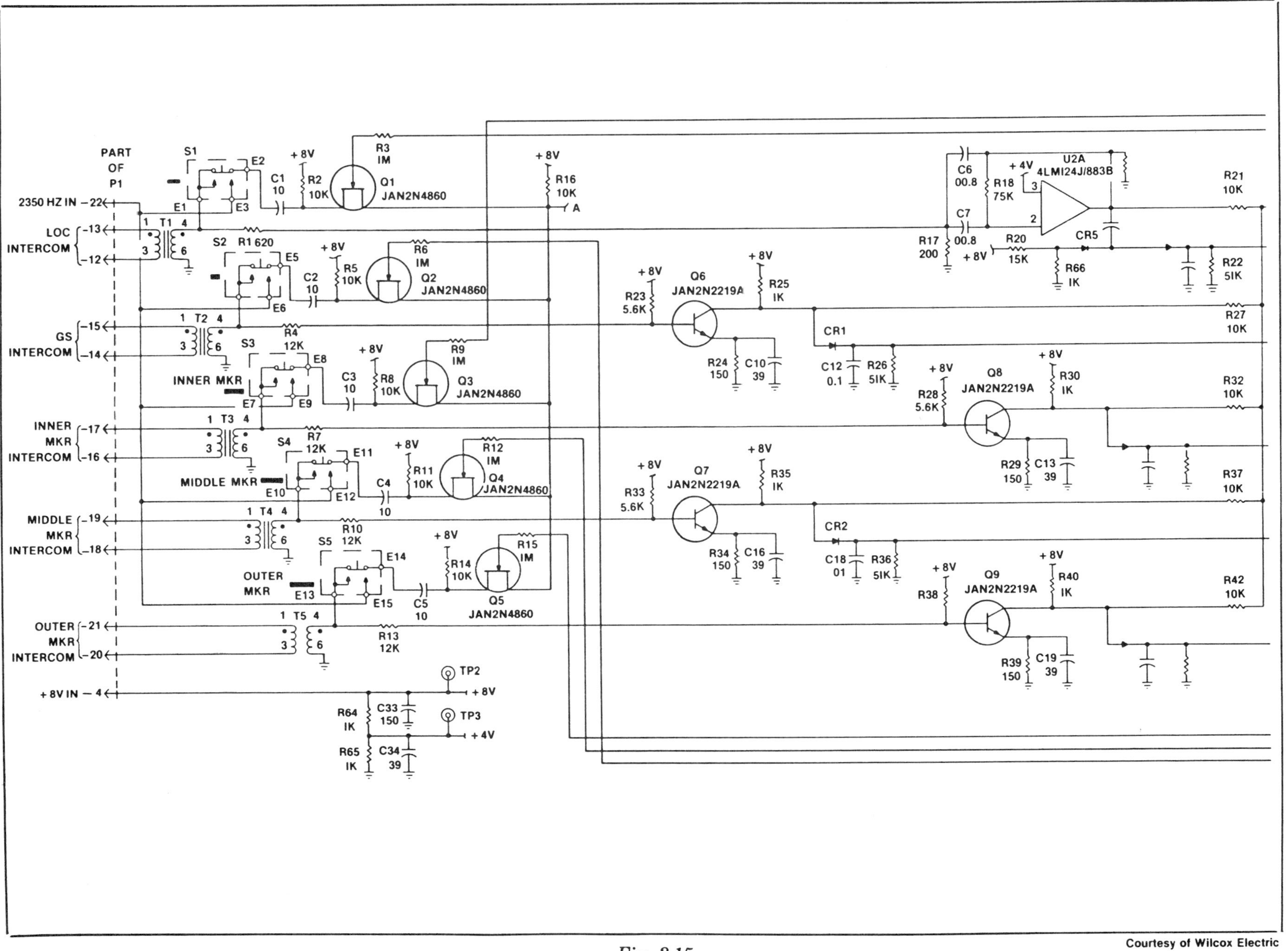

Fig. 3-15

Courtesy of Wilcox Electric

SECTION IV

Resistance and Ohm's Law

Although Section 3 discussed the diagrams used to describe circuits and defined the characteristics of both open and closed circuits, most of the section was devoted to an explanation of how DC sources are arranged. In this section, we will begin to deal with more complicated circuits, basing our work on the generalized concept of a *load*, an energy-absorbing or energy-using device of any sort connected to a current source. In our discussion of DC circuits, loads will be limited to circuit elements that impede the flow of an electric current, in other words, circuit elements that possess *resistance*. As we shall see, the potential difference across a load, its resistance, and the current passing through it, are related by a simple mathematical formula called *Ohm's law*.

In working out practical problems involving Ohm's law we often meet with high resistances and tiny currents, on the order of 0.001 ampere (10^{-3} A) or less. For this reason it is necessary to begin this section with a summary of the prefixes used to denote very large and very small quantities.

A. Measurement Prefixes

The familiar English measurement system includes a great many different units whose interrelationship often appears haphazard. For example, there are twelve inches in a foot, three feet in a yard, and 1760 yards in a mile, all of which are in no way obvious from the names of the various units. The matter becomes even more confusing when we deal with weight or volume measurements. For the European, however, who grew up with the metric system, the matter is greatly simplified. Once a very few simple numerical prefixes are memorized, the rest is easy. For many years, the scientific community all over the world has been using these prefixes, which are based on multiples of 10. Table 4-1 gives the most common of these prefixes along with their numerical values and examples of their use.

At first glance, the material presented in Table 4-1 may seem overwhelming, but it should be studied thoroughly. In the course of a day's work, the avionics technician will be constantly bombarded by these prefixes, and his response to such questions as how many microfarads are represented by a 1100-picofarad capacitor must be automatic. In the light of this requirement, a large number of prefix exercises have been included among the self-test questions.

B. Circuits and Loads

As was briefly suggested previously, an electric field which is driving a current through a wire or other circuit is actually doing work, since the materials composing the circuit resist the passage of conduction electrons. In the main, this resistance to electron flow is due to the actual blocking of conduction electron drift by the atoms of the conducting wire or circuit. This is illustrated by the effect of temperature upon the resistance of most common metals whose conduction electrons form a "soup" as described in Section 1.

When the temperature of the conducting metal is lowered, the resistance decreases, actually disappearing completely at very low temperatures. This is because the vibration of the atoms in the metal crystal lattice ceases at very low temperatures. A stationary atom in a crystal lattice presents a very small cross section to the drifting conduction electrons, and very few of the drifting electrons are captured by the metal atoms. However, if the temperature of the metal is increased, the added heat energy causes the atoms in the lattice to vibrate. This means that the metal atom is more likely to "get in the way" of a drifting conduction electron and to capture it.

In addition, even though the drift velocity of the conduction electrons due to the electric field remains fairly constant, the overall velocity of any individual electron is increased by the available

PREFIX	ABBRE-VIATION	DECIMAL VALUE	EXPONENTIAL VALUE	EXAMPLES OF USE
pico- (formerly micromicro ($\mu\mu$) in some usages)	p	0.000000000001	10^{-12}	one picosecond = 10^{-12} seconds; 270 picofarads = 0.00000000027 farad, a typical capacitor value
nano-	n	0.000000001	10^{-9}	one nanofarad = 0.000000001 farad, another very common value for a capacitor; 250 nanoseconds = 0.00000025 seconds, the time it takes for a fairly fast computer memory device to look up a stored value
micro-	μ	0.000001	10^{-6}	three microvolts = 0.000003 volt, a typical potential difference caused in a radio receiver antenna by a weak signal
milli-	m	0.001	10^{-3}	ten milliamperes = 0.01 amp, the current required to light a light-emitting diode (LED) device such as that used in place of indicator bulbs in modern electronic equipment
centi-	c	0.01	10^{-2}	the familiar centimeter, 0.01 meter
deci-	d	0.1	10^{-1}	rare in electronics; the decibel, 0.1 bell, a comman way of expressing power or voltage ratios
deka-	D	10	10^{1}	rarely used, but familiar in the word decade, e.g., a decade or ten-position switch
kilo-	K	1000	10^{3}	one of the most widely used of the prefixes; 10 kilovolts = 10,000 volts; 3 kilowatts = 3000 watts; 100 kilohm = 100,000 ohms
mega-	M	1,000,000	10^{6}	ten megohms = 10,000,000 ohms, a very high but still commonly found resistance; 328 megahertz = 328,000,000 cycles per second, the approximate frequency of the glide slope portion of an instrument landing system
giga-	G	1,000,000,000	10^{9}	prefix generally used only to express a very large number of cycles per second (hertz); 3 gigahertz = 3,000,000,000 cycles per second

TABLE 4-1

thermal energy when the temperature of a conducting wire increases. This means that although a conduction electron will gradually drift toward the positive pole of the battery connected across the wire, it will not do so directly. Instead, it will rush about in a random manner, colliding with the vibrating atoms of the crystal lattice. The more a conduction electron rushes about, the greater the chance that it will be captured by an atom.

As discussed in Section 1, when a conduction electron is captured, it falls into an outer orbit around the atom which has captured it. When this occurs, it gives up some energy in the form of radiation. This liberated energy may be in the form of heat, and if so, serves to raise the temperature of the conductor farther. Ultimately, the conduction electron will absorb enough energy from the electric field to free itself from the capturing atom, repeating the process. The effect then is a conversion of the energy of the electric field into heat, which in itself is a form of energy.

The energy which is turned into heat when a current passes through a conductor is obtained from the source. At the very lowest temperature possible, in the neighborhood of −273 °C, another effect is observed. Atomic vibration ceases, and the conduction electrons possess only the drift velocity. In this case, the capture of a conduction electron by conductor atoms is rare and requires almost no energy from the source to produce a high current. This effect is known as *superconduction*. But, because such low temperatures are required to bring about superconduction, the effect can be used for only a few special applications.

Thus, whenever a current passes through a conductor, energy from the electric field is converted into heat. Since the energy possessed by an electric field, at a point, is proportional to the voltage difference between that point and a reference point, (generally the negative pole of the current source) we might expect that the voltage at various points along a closed circuit will decrease the farther we get from the positive pole of the current source and the closer we get to the negative pole or reference point.

Fig. 4-1 shows a uniform wire, ten meters long, connected across a battery. The battery provides a potential difference of 6 volts across the ends of the wire. Since the wire represents a closed circuit, a current will begin to flow. A certain amount of electrical energy will be converted into heat, thereby decreasing the potential difference between each successive point along the wire and the reference point at the negative pole of the battery. At point B, exactly halfway along the wire, the potential with respect to the reference point will be 5/10 × 6 volts, or 3 volts. Therefore, we may say that one-half the potential difference or voltage furnished by the battery is used up in driving a current from the positive terminal of the battery to point B. Remember that this statement is made from the point of view of current flow, although the physical flow of electrons is in the opposite direction.

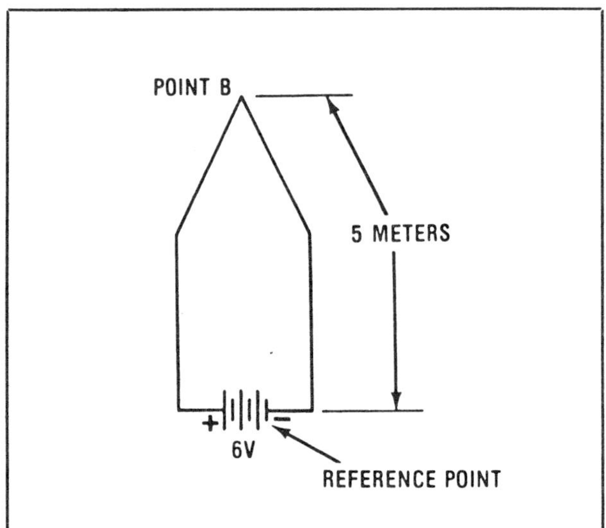

Fig. 4-1 Potential difference along a uniform wire connected across a battery.

In Fig. 4-1, the ten-meter-long wire will be heated because of the passage of the current through it. Since energy supplied by the battery is used in this process, the wire may be considered as a load.

The opposition to current flow exhibited by this load is not confined to a particular point, but is distributed along the entire ten-meter length of the wire. This is why it is referred to as a *distributed load.* For the purposes of drawing schematic diagrams, it is highly inconvenient to attempt to show a distributed load, and so, for the sake of calculation and ease of drawing, such a load is considered as though all the opposition to the flow of current existed at only one point. It is represented, in general, by a small rectangle or zig-zag line as shown in Fig. 4-2.

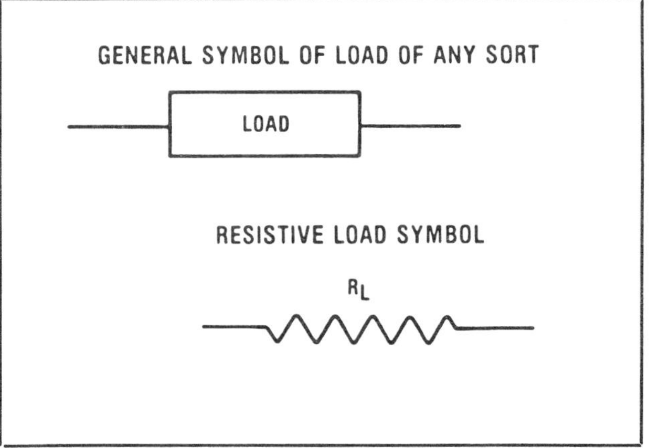

Fig. 4-2 Symbols for general load and resistive load used in schematic diagrams.

The wires connecting the load to the battery are assumed then to exhibit no opposition to the flow of current. Instead, all opposition, or resistance, is combined, or *lumped* into one load. Such a representation therefore considers the wire as though it were a *lumped load.* A *true* lumped load, for example, could be a light bulb with a fairly high resistance to current flow connected to a battery by heavy wires showing a relatively small resistance. In most DC circuits, it is relatively safe to deal with distributed loads as though they were lumped, although this is not always the case in AC circuits.

1. Linear and non-linear loads

In the preceding example, the 6-volt battery was found to produce a current of 1 ampere through the load represented by 10 meters of a particular copper wire. If we substitute a 12-volt battery for the 6-volt one, we would observe a current of 2 A through the wire. Thus by doubling the voltage

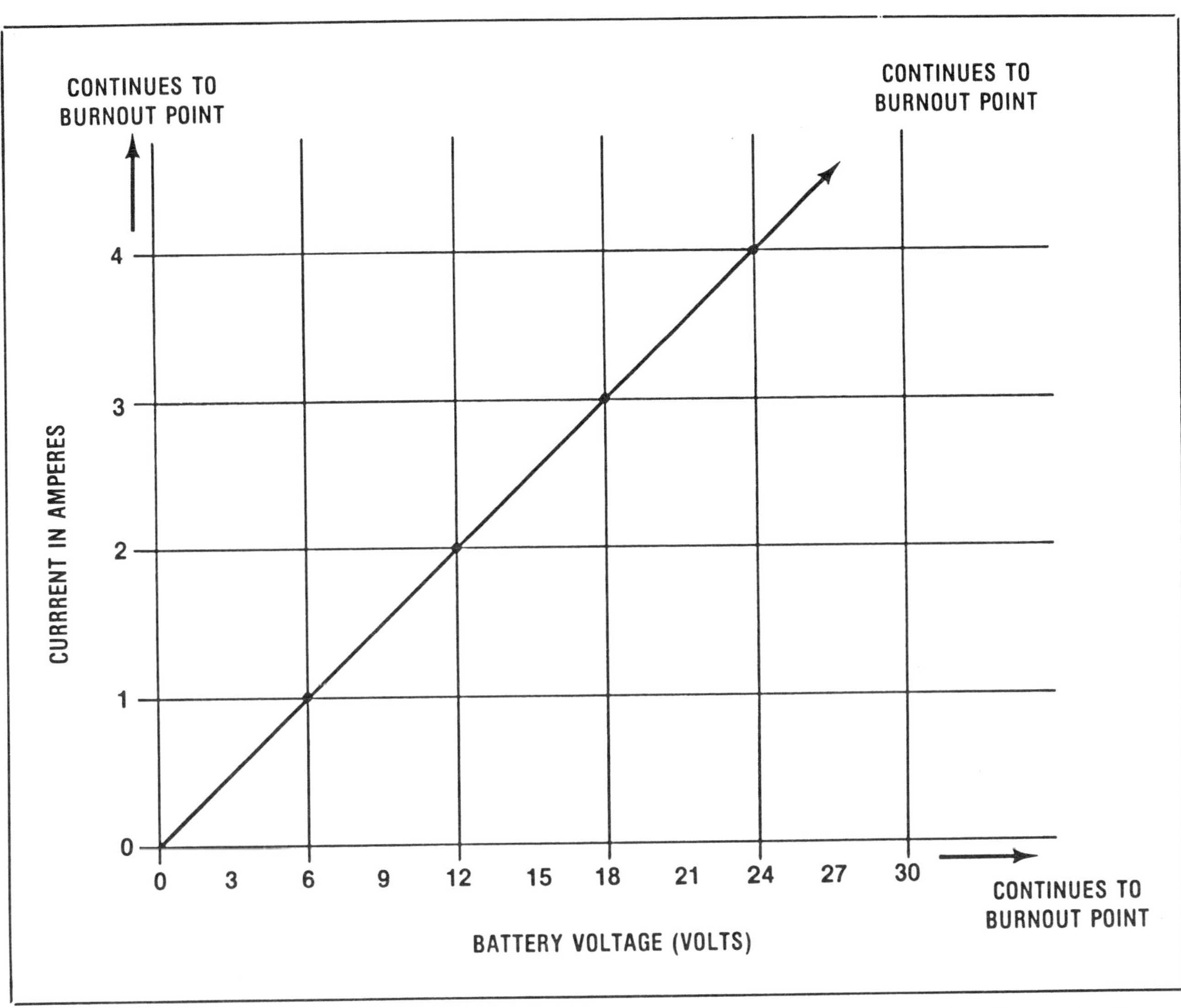

Fig. 4-3 Graph of the current against the potential difference across a resistive (linear) load.

across the load, we double the current through it. Substituting a 3-volt battery for the 12-volt battery would reduce the current through the load to 0.5 A.

If we try various batteries, we may draw a graph of the battery voltage and load current as shown in Fig. 4-3. Notice that the graph line which represents all the possible currents for all the possible voltages is a straight line, limited only by the 0 volts — 0 A point and a point out to the right where the wire would burn out.

A material such as copper which shows a straight line current versus voltage curve is said to exhibit a *linear resistance curve*. This means that multiplying the voltage by any number, say 2.3, will result in multiplying the observed current by the same number. In reality, no materials have a completely linear resistance curve, as increasing the temperature will increase the resistance of most common conductors. In some cases opposition to the flow of current drops sharply when the voltage passes a certain point. This occurs in ionized liquids. In other materials, the heating effect caused by the passage of large currents makes it increasingly difficult to produce any increase in the current flow. In general, the materials with which the avionics technician is dealing are chosen to retain their linearity (assuming that they are supposed to be linear in the first place), over the range of voltages and currents normally used.

This does not mean that non-linear devices are useless. Quite the contrary. For example, the device called a current amplifier allows a small increase in voltage to produce a large increase in the current through a circuit. The opposite of the current amplifer effect, as in the transient suppressor, is frequently used to keep the current through a circuit fairly constant in spite of an increase in the applied voltage. We use the transient suppressor to protect sensitive electronic equipment from the short-term increase in powerline voltages caused by nearby lightning strikes. But, in order to calculate the current in any circuit from the applied voltage, our linear resistance loads must be as linear as possible and our non-linear loads as predictable as possible.

2. *Ohm's law*

The relationship between the voltage applied to a linear load and the current through it was discovered by the German physicist George Simon Ohm about 1830. He found that for a particular length and thickness of a linear resistance conductor, the ratio of the applied voltage to the current through the conductor remained constant, regardless of changes in the applied voltage. In other words:

$$\frac{\text{applied voltage}}{\text{current}} = \frac{E}{I} = K \text{ (a constant).}$$

This constant is the *resistance* of the conductor. It is measured in ohms, with one ohm being the resistance through which a potential difference of one volt will drive a current of one ampere. In its final form then, Ohm's law may be mathematically stated as:

$$R \text{ (ohms)} = \frac{E \text{ (volts)}}{I \text{ (amperes)}}$$

(Equation 4-1)

In words, the resistance of a load or circuit in ohms is equal to the voltage applied across the load or circuit in volts divided by the current through it in amperes. By simple algebra, this equation may be converted into:

$$E \text{ (volts)} = I \text{ (amperes)} \times R \text{ (ohms)}$$

(Equation 4-2)

This equation states that the potential difference or voltage across a circuit or a load is equal to the product of the current through the load in amperes, and the resistance of the circuit or load in ohms. A third and final form of Equation 4-1 is:

$$I \text{ (amperes)} = \frac{E \text{ (volts)}}{R \text{ (ohms}}$$

(Equation 4-3)

This equation states that the current in amperes through a circuit or load is equal to the applied voltage in volts divided by the resistance in ohms. These three ways of stating Ohm's law should be firmly kept in mind, for they represent three different forms of the most frequently used mathematical relationship in electronics.

These forms of Ohm's law are so important that a great many little tricks have been thought up to aid in remembering them. The author never had any luck with such tricks, and generally recommends memorizing the form $E = I \times R$ at first, since the two other forms may easily be derived from it by a simple algebraic process.

a. *Examples of Ohm's law calculations*

Example 1:

A load with a resistance of 10Ω (ohms) is connected across a source of unknown voltage. The current flowing through the load is found to be 500 milliamperes. Assuming that the wires from the source to the load are without resistance, what is the voltage of the source?

Answer:

Using the form of Ohm's law $E = I \times R$

$E = 0.5 \times 10$ (recall that 500 milliamperes = .5 ampere)

$$E = 5 \text{ volts}$$

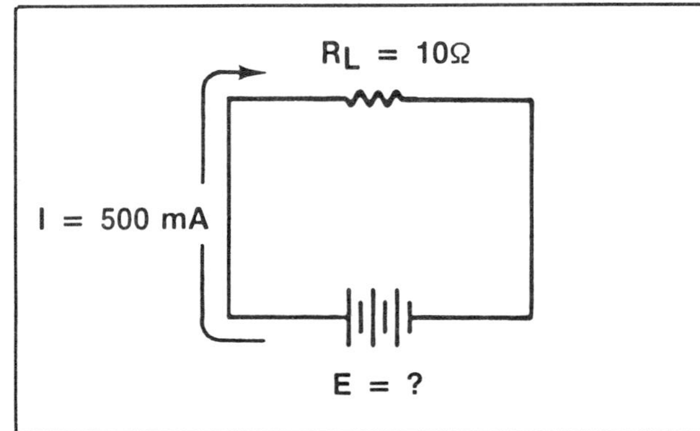

Fig. 4-4 Circuit for examples 1 and 2 of Ohm's law calculations.

Example 2:

The load of Fig. 4-4 is increased to 1 k ohm; what is the resulting current through the load?

Answer:

In this case the voltage E of the source remains 5 volts, but the resistance is now 1 k ohm (1000 ohms). We use the formula

$$I = \frac{E}{R} \text{ or } I = \frac{5}{1000} = 0.005 \text{ A, or}$$

5 milliamperes (mA).

Example 3:

A battery whose potential difference is 9 volts is shown in Fig. 4-5. The battery is capable of furnishing a maximum current of 100 mA without damage. Ignoring the resistance of the wires connecting the battery to the load, calculate the minimum load resistance possible for this battery. What would be the current through the load if the load resistance were half this value?

Answer:

Use the form of Ohm's law:

a. $R = \frac{E}{I} = 9 \text{ volts and } I = 0.1 \text{ A then}$

$$R = \frac{9}{0.1} = 90 \text{ ohms}$$

b. If the resistance is halved, we might expect the current to double, or to be equal to 200 mA

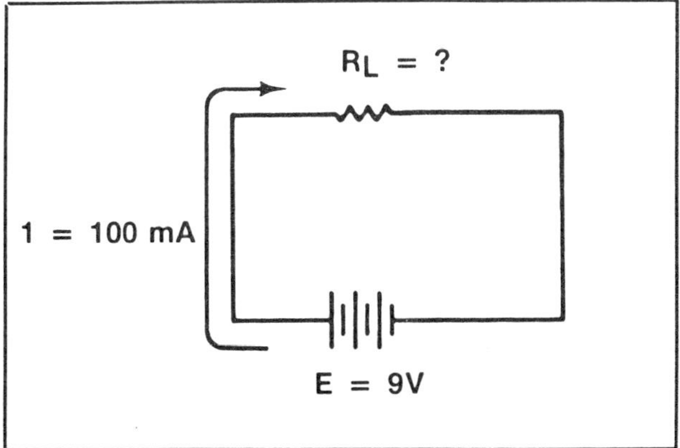

Fig. 4-5 Circuit example of an Ohm's law calculation.

(0.2 ampere, or twice the maximum allowable battery current). Let us check this using the form of Ohm's law:

$$I = \frac{E}{R} = \frac{9}{45} = 0.2 \text{ ampere}$$

3. Loads in series

In the previous examples, we have assumed that all the resistance in each circuit was lumped in the load, and that the wires used to connect the loads to their current sources were without resistance. Of course, this is not true. As we saw previously, only in the state of supercondition can a wire or any other conductor be without resistance. Practically, however, the resistance of connecting wires used in electronic circuits is considerably lower than the resistance of the loads involved. Therefore, since they possess less than one per cent of the resistance of the circuit, we can usually ignore them.

In electric power distribution circuits this is not the case. Here, wire resistance must be taken into account, generally as one or more additional loads in series with the load connected to the wires. In Fig. 4-6, load B represents some energy-using device like a lamp or heating coil, while loads A and C represent the resistances of the wires connecting the source to load B. If A and C were not present (i.e., were zero ohms) the current through

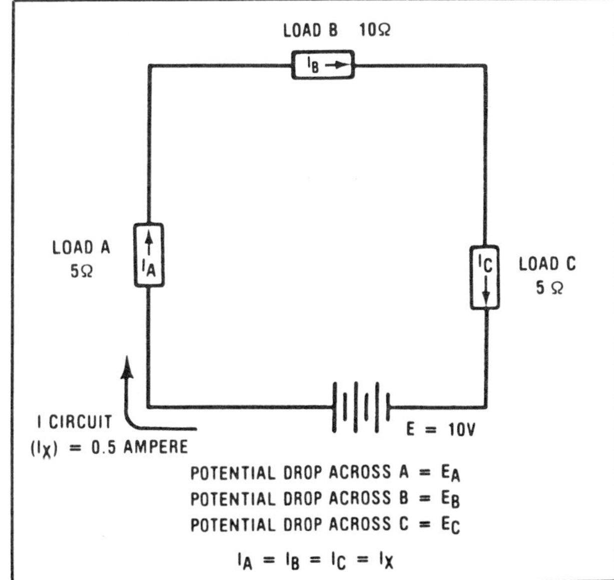

Fig. 4-6 Circuit showing the loading effect of connecting wire resistance.

load B would be 1 ampere. But this cannot be the case in the circuit of Fig. 4-6, because the 10-volt potential difference of the current source must now drive the current through *three* loads. At this point, let us use a bit of ingenuity and Ohm's law to calculate the resistance of this series circuit.

Obviously, any current that flows through A must also flow through B and C, since there are no branches or alternate routes. Let us call this current I_X. The potential drop across a load by the first form of Ohm's law is equal to the current through the load times the resistance of the load or $E_{drop} = I_{load}R_{load}$. For load A, the voltage drop $= I_X \times 5 = 5I_X$. For load B, the voltage drop $= I_X \times 10 = 10I_X$. For load C, the voltage drop $= I_X \times 5 = 5I_X$. The total voltage drop across all the loads in the circuit is then equal to $20I_X$. But we know that the voltage across the entire circuit is equal to the source voltage or 10 volts. Since there is no other source of voltage in the circuit, the total voltage drop must be equal to 10 volts, and so $20I_X = 10$. I_X is then equal to 0.5 A, which is the current through the circuit composed of the three loads. But this is exactly the result that would be obtained if the three load resistances were simply added (5 + 10 + 5 = 20) and substituted in

$$I = \frac{E}{R} = \frac{10}{20} = 0.5 \text{ A.}$$

The result of this demonstration shows that the resistances of series-connected loads may be simply added to determine the total resistance of a circuit. This total resistance may then be considered as a distributed resistance which has been "lumped" for purposes of calculation in Ohm's law equations.

Thus, the total resistance of the circuit shown in Fig. 4-7A is the sum of the resistances of loads A and B. The current through both loads is the same since they are connected in series. The resistance of the entire circuit is $R_A + R_B$, or 10.1 ohms. (Note: schematic diagrams frequently use the symbol Ω, the Greek letter *omega*, to represent ohms.) The current I through both loads will be

$$\frac{E}{R_A + R_B} \text{ or } \frac{6}{10.1} = 0.594 \text{ A.}$$

The voltage drop across load A can then be determined by using $E_A = I \times R_A$, or 0.0594 volt, while the voltage drop across B, E_B, is equal to $I \times R_B$ or 5.94 volts. As a check, it may be noted that $E_A + E_B$ is equal to 6 volts when rounded off, showing that the total of the voltage drops across the loads in a circuit is equal to the total potential difference supplied by the source. It is important to note that the voltage drop across load A, which represents the relatively low-resistance hook-up wire, is small as compared to the voltage drop across load B.

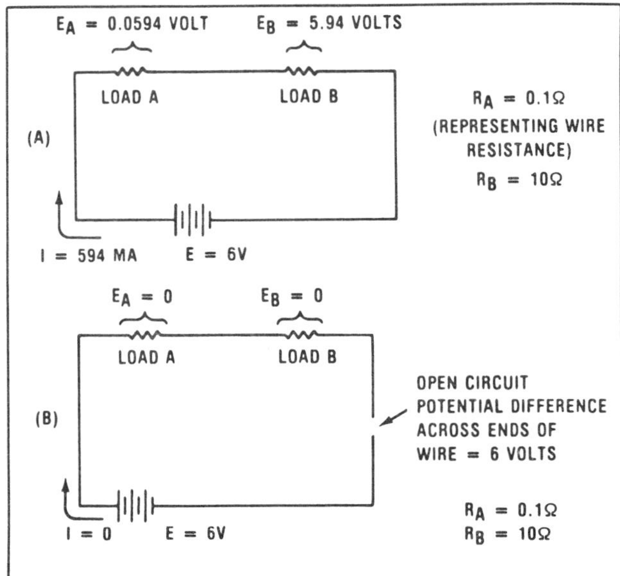

Fig. 4-7 A series circuit A — the current through both loads is the same. B — Broken at one point. Notice that the current through the loads drops to zero.

In Fig. 4-7B the circuit shown has a break in the wire at one point. In this case, the circuit current is zero since the current path is broken at the open circuit point. The resistance of the two loads remains the same, since it is based on the physical structure of the loads themselves. For linear resistances the load resistance is independent of the current. The voltage drop across the two loads is zero since the current is zero and the voltage drop is equal to the current times the resistance ($I \times R$). The potential difference across the gap (as explained in the discussion of open circuits) is the same as that of the battery since there are no voltage drops in the circuit.

We may look at Fig. 4-7B in another way, by viewing the break in the circuit not as an actual break, but as a load with a very high resistance. If this resistance is great enough (on the order of one thousand times the sum of the other loads in the circuit), nearly the whole of the potential difference supplied by the source will appear across the high resistance load.

Fig. 4-8A pictures a circuit of this sort, comparing it with a circuit (4-8B), in which the loads are more nearly equal. Should one of the loads in the circuit of Fig. 4-8B burn out, creating an open circuit, this would be quite easy to detect using only a voltmeter. Measuring the voltage drop across each of the loads would show a zero drop across the good load and a voltage drop equal to the source voltage across the burned-out load. If both loads are good, on the other hand, a voltage drop equal to I × R will be measured across each of the loads. This checking procedure is quite a common one, but will not work if there is a great difference between the values of the series loads as in Fig. 4-8A, where the burning out of the high resistance load would be hard to detect by measuring the voltage drops with the instruments in general use by avionics technicians.

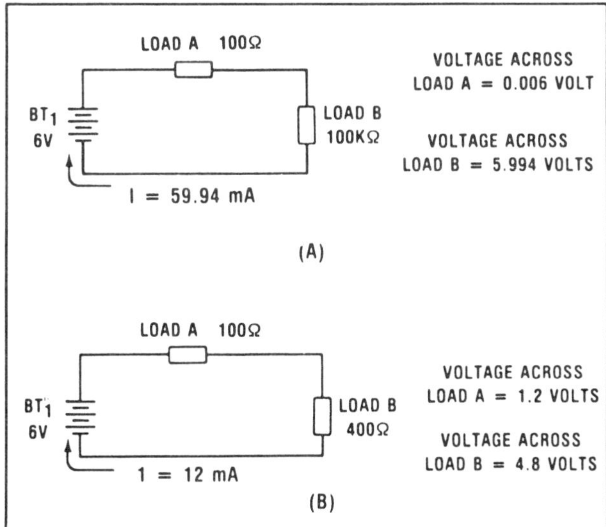

Fig. 4-8 Loads in series — measuring potential differences across loads can serve as an aid to troubleshooting.

4. *Current limiting loads*

The lower resistance load in a series combination of a moderate and a high resistance load has little effect on either the circuit current or on the voltage drop across the high resistance load. Therefore, it might be thought that such an arrangement would not be much used in practice. Actually, this arrangement is very common if the high resistance load is the output of a digital logic device or a transistor switch.

In Fig. 4-9 such an arrangement is pictured. The high resistance load is replaced by a *two-state device*. This is a component which may be switched from a high resistance to low resistance condition by a control signal. When a two-state device is in its high resistance condition, we want almost all the potential difference supplied by the source to appear across the device.

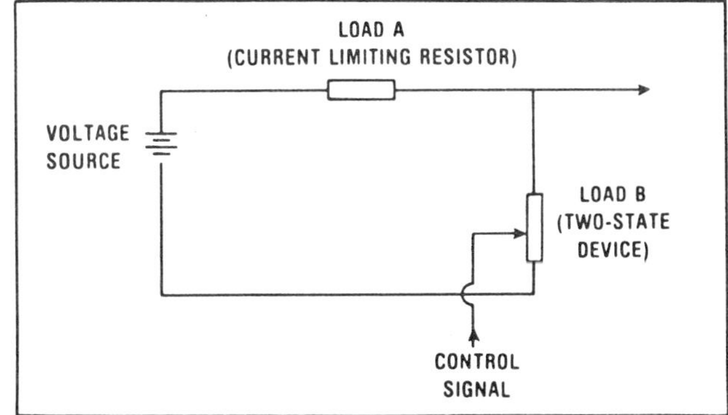

Fig. 4-9 Current limiting resistors.

Therefore, the load in series with it (load A), must not have a resistance high enough to cause an appreciable voltage drop (I × R). On the other hand, when the two-state device is in its low resistance condition, the amount of current through the circuit is almost entirely dependent upon the resistance of load A. Since most modern electronic two-state devices such as logic elements and transistor switches cannot tolerate high currents, it is the job of this load to provide enough circuit resistance to keep the current down to a safe level. For this reason, load A resistance must be high enough to limit the series circuit current. Fig. 4-10 pictures a schematic showing a number of resistors which function in this way.

5. *Voltage divider circuits*

One of the most common applications of series-connected loads is to provide a particular potential difference or a number of potential differences from one voltage source. Fig. 4-11 shows a 12-volt source connected to two loads in such a way as to provide a potential difference of three volts be-

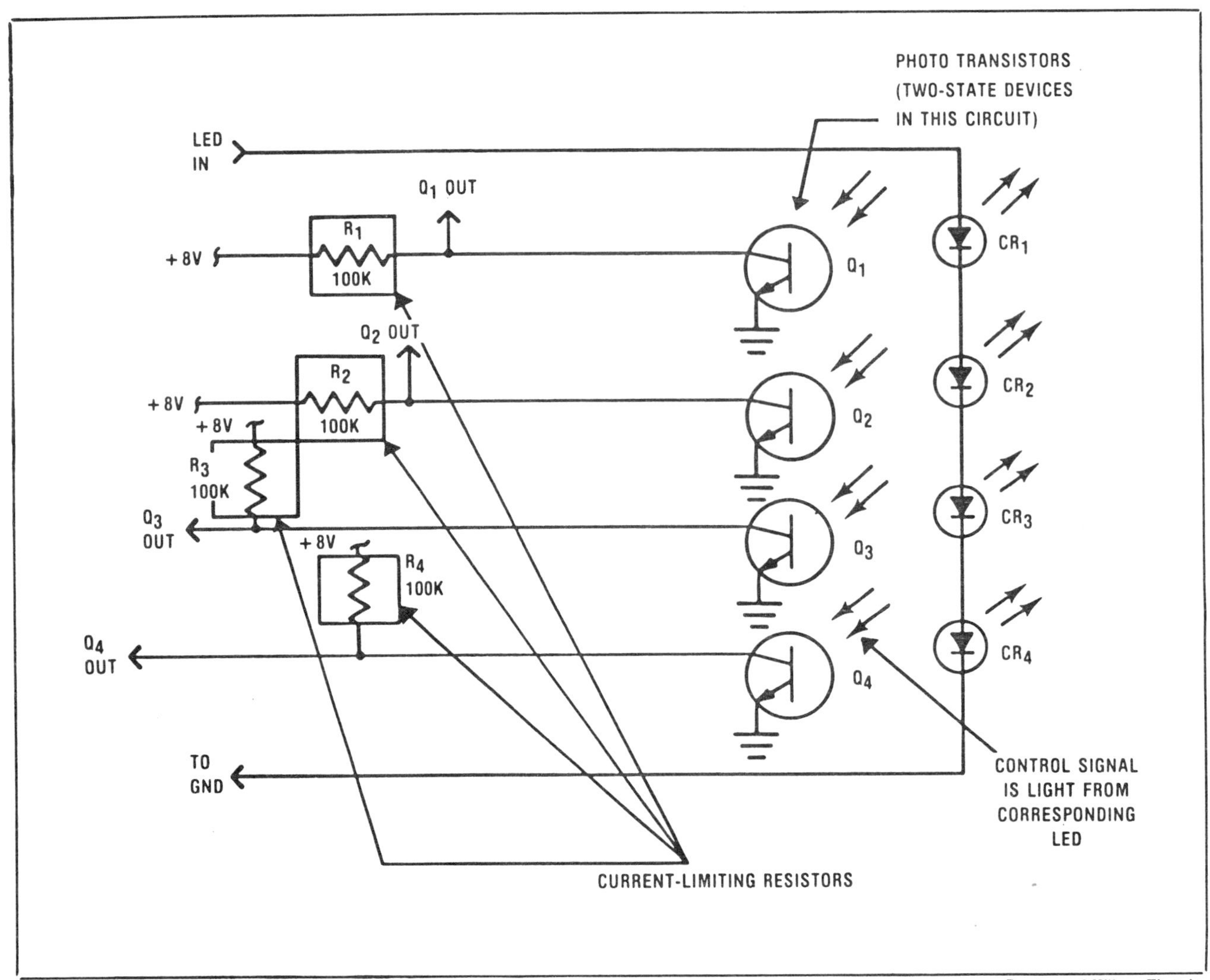

Courtesy of Wilcox Electric

Fig. 4-10 Current-limiting resistors — example taken from an operating tilt detector.

BT_1
12 VOLT
LOAD A 30Ω
OUTPUT 3 VOLTS
LOAD B
10Ω
I = 0.3 AMPERE
GROUND CONNECTION

Fig. 4-11 A simple voltage divider — note that this circuit is not required to furnish a current to an external load.

tween the output terminal and the reference terminal (in this case, as in most electronic equipment, a general reference point called the *ground* is established and is usually connected at one point to the chassis and cabinet of the unit and to the ground line of the main power distribution system.) The voltage across a load in a divider may be calculated by forming a proportion:

$$\frac{\text{E across section}}{\text{E total}} = \frac{\text{R section}}{\text{R total}}$$

(Equation 4-4)

This may also be written as:

$$\text{E across section} = \frac{\text{R section}}{\text{R total}} \times \text{E total}$$

(Equation 4-5)

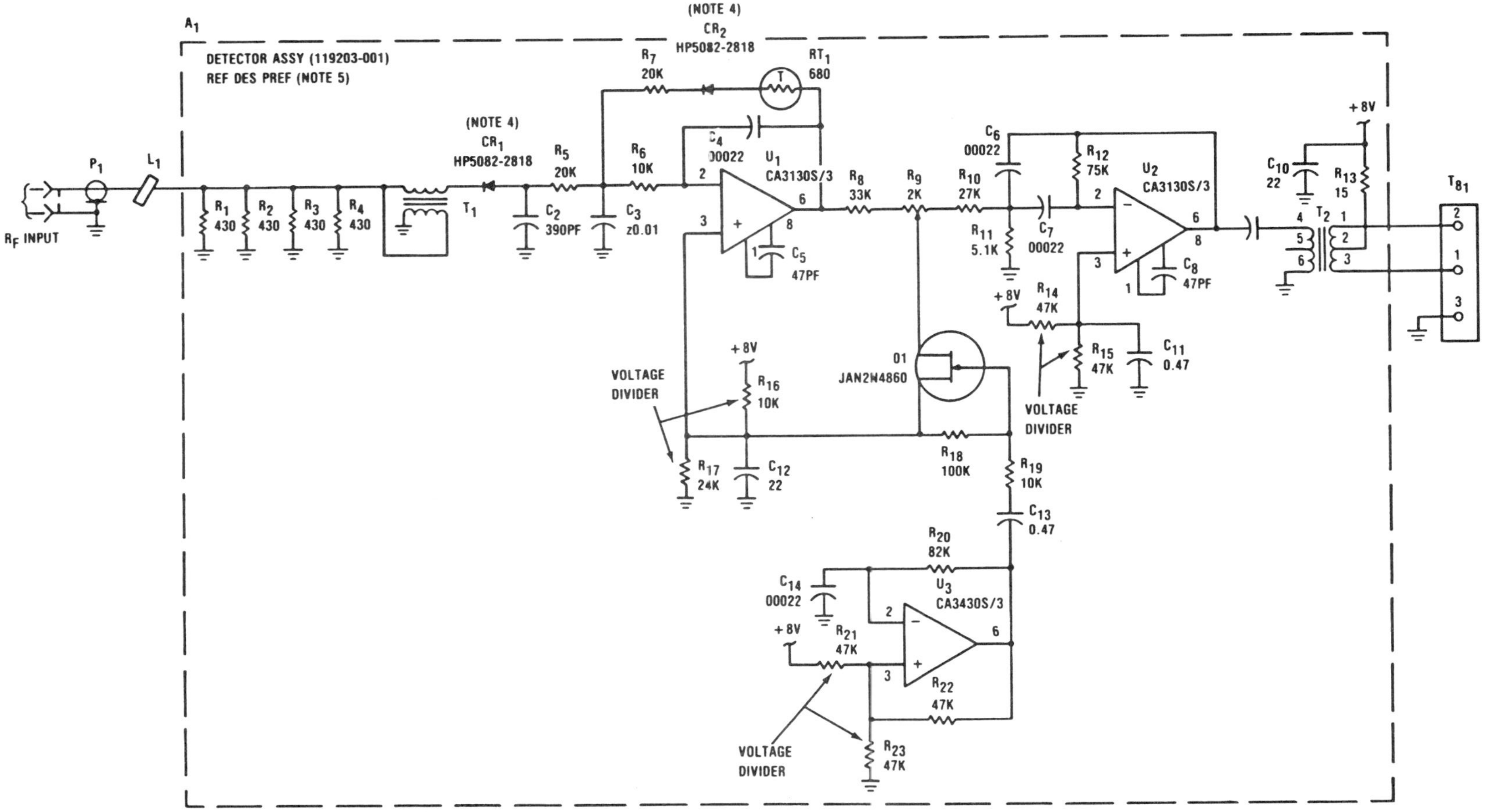

NOTES:

1. VALUES ARE IN OHMS OR MICROFARADS UNLESS OTHERWISE INDICATED.

2. DENOTES LABELED ITEM.

3. I.C. PIN CONNECTIONS NOT SHOWN: PIN 7(V_{CC}) OR U_1, U_2, AND U_3 TO +8V; PIN 14(V_{SS}) TO GND.

4. DIODES CR_1 AND CR_2 ARE A MATCHED PAIR.

5. PREFIX REFERENCE DESIGNATORS WITH $LOCA_1/A_2$ OR GSA_1 (NR AND SBR GS STATIONS).

Courtesy of Wilcox Electric.

Fig. 4-12 Practical voltage dividers.

These equations produce only the ratio between the load across which the voltage is derived and the total resistance of the circuit, 1:4 in the example of Fig. 4-11. These equations are meaningful only if the divider is not required to furnish current to some external circuit, unlike the practical voltage dividers shown in Fig. 4-12.

In a later section, the principles required for the actual design of a practical voltage divider system will be covered. The importance of this cannot be stressed highly enough, for one of the basic jobs of an avionics technician, the installation of new electronic equipment in older aircraft, or ground installations, will frequently involve matching the power requirements of the equipment to the power available in the aircraft or at the site.

Voltage dividers are not limited to supplying only one voltage. In Fig. 4-13, we have a voltage divider formed of four loads which, working from a 12-volt source, is able to furnish 12-, 9-, 6-, and 3-volt potential differences with respect to a single reference point (ground).

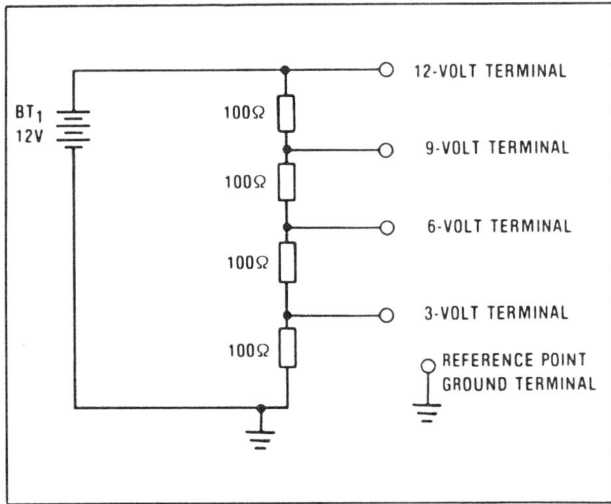

Fig. 4-13 Voltage divider designed to provide reference potential differences; it can not furnish current to external loads.

A. Self-Test Questions:

Complete the following using exponential notation for numbers greater than 1000 or less than 0.001.

1. One ________ ampere is another way of writing 0.001 ampere.
2. The symbol Ω is the usual abbreviation for ________.
3. Ten thousand herz is normally written as ten ________ herz.
4. 1 MΩ is short for ________ ________ ________.
5. One nanofarad = ________ picofarads = ________ microfarads.
6. Micromicrofarad, abbreviated μμf, is the older form of ________.
7. Ten gigaherz (GHz) = ________ megaherz (MHz) = ________ herz (Hz).
8. 18 milliamperes (18 mA) = ________ nanoamperes (nA) = ________ microamperes (μA).
9. 33.6 μvolt = 0.________ volts.
10. One microvolt = ________ kilovolts.
11. Twelve picofarads = ________ nanofarad.
12. Thirty centimeters (cm) = ________ millimeters = ________ meters.
13. If a ten-volt source is connected across a 1.5 k ohm load, the circuit current will be ________ milliamperes or ________ amperes.
14. Two 30Ω loads and a 50Ω load are connected in series across a battery formed by six 1.5 volt cells connected in series aiding, the resulting current supplied by the battery is ________ mA. The voltage drop across the 50Ω load is ________.
15. When a 15-volt source is connected across a load, the resulting current is 33 milliamperes; the resistance of load is ________ ohms.

B. Questions for Thought:

1. a. Calculate the total current through and the voltage drop across load A of Fig. 4-14.

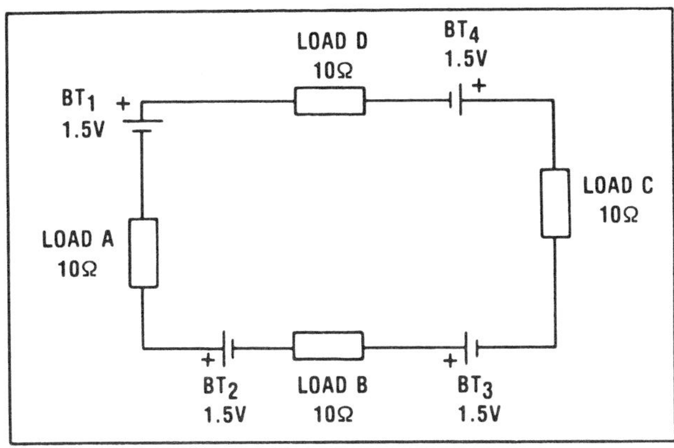

Fig. 4-14

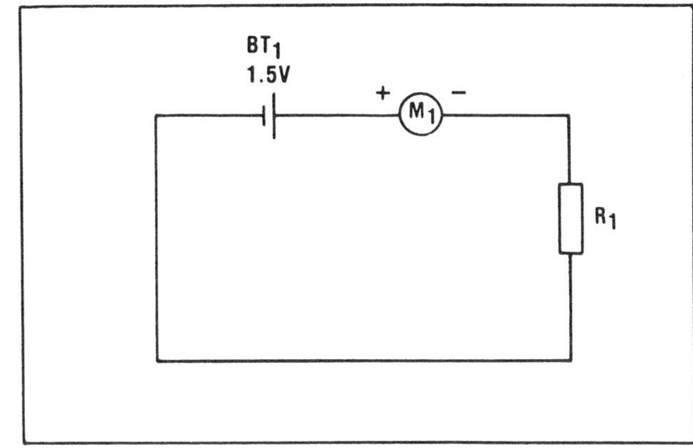

Fig. 4-15

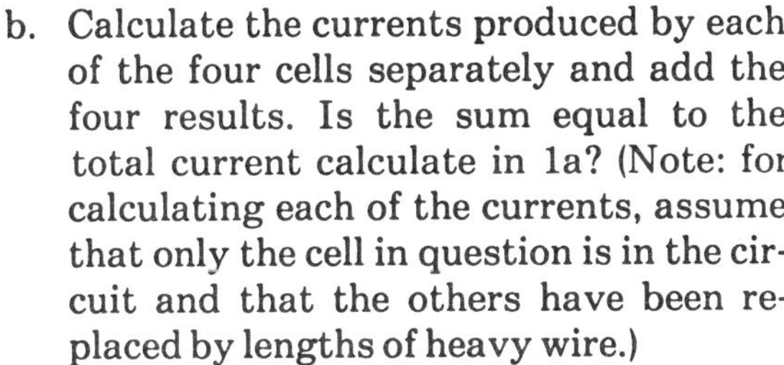

b. Calculate the currents produced by each of the four cells separately and add the four results. Is the sum equal to the total current calculate in 1a? (Note: for calculating each of the currents, assume that only the cell in question is in the circuit and that the others have been replaced by lengths of heavy wire.)

c. Try to state the principle of this exercise in a general way. (We will see that the answer to this question is one of the basic ideas of network analysis, the general method of determining what is going on in complex electronic circuits.)

2. Name as many electrical loads as you can; for each, give the form of energy or work into which the load converts electrical energy.

3. It was shown that a break in a wire may be considered as a load of very high resistance. Is it a linear or non-linear resistance? Justify your answer.

4. Try to find as many voltage dividers as you can in the schematic diagram of Fig. 3-15.

C. *Experiments and Demonstrations:*

1. Using a multimeter (M_1) with a current measuring capacity of at least 1 A, a ten-ohm, 20- or 25-watt resistor (R_1) and four 1.5-volt D cells ($BT_1 - BT_4$), construct the circuit shown in Fig. 4-15.

   a. Record the voltage of the cell and the current through the circuit.

   b. Adding one cell in series at a time to aid the circuit, record the total source voltage and the current through R_1. Draw a graph of your results like the one shown in Fig. 4-3.

   c. Does Ohm's law hold for the resistor?

   d. Perform the same experiment using a six-volt bulb in series with R_1.

2. Given a battery composed of up to four 1.5-volt D cells connected in series, parallel or series parallel, and up to six 10-ohm loads, design and build a power supply (battery plus voltage divider) that will furnish:

   a. 0.75 volt

   b. 4.5 volts, 2.25 volts, and 3 volts

   c. 3.6 volts, 2.4 volts, and 1.2 volts.

SECTION V

Parallel Loads and Power in Loads

In this section the application of Ohm's law is extended to include loads connected in parallel. This type of circuit requires a different treatment from the series connection discussed in the last section.

In Section 4, it was pointed out that the series circuit contains no branches and that the current flowing through any point is *exactly* the same as the current through any other point in the circuit. This is not the case in the parallel circuit, where at least two branch paths are available. Since highly complex DC circuits are quite common in avionics equipment, the first part of this section is designed to enable the avionics technician to calculate the voltage drop across and the current through loads in parallel and series-parallel.

The concept of an *equivalent circuit*, a simple circuit that possesses the same overall characteristics as a more complicated one, is introduced to help calculate current through and voltage drop across branch circuits. The last part of this section introduces the notion of power in electronic circuits. Power loss in loads and the effect of power upon the choice of loads in practical circuits are also introduced.

A. *Parallel Loads*

The discussion of Ohm's law in series circuits in Section 4 stressed the fact that the current through all parts of the series circuit is the same. This is understandable, since there is only one possible current path from the positive terminal to the negative terminal of the current source in a series circuit as shown in Fig. 5-1A. In a parallel circuit (Fig. 5-1B) there are two possible paths from the positive pole of BT_1 to the negative pole, one through load A, the second through load B. From the illustration, one might guess that the current through load A will not be the same as the current through load B, since B represents a "harder" path than A, the resistance of B being twice that of A.

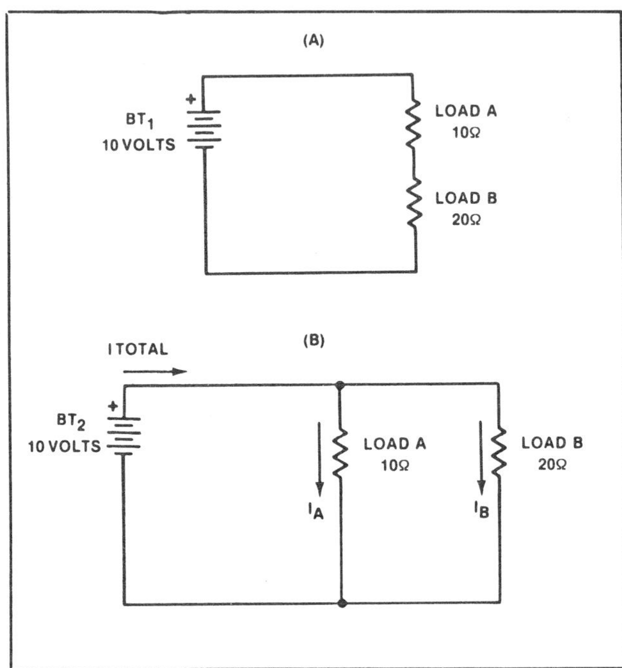

Fig. 5-1 Illustration of the differences between series and parallel circuits. A parallel circuit consists of two or more branches.

On the other hand, the potential difference across both loads is the same since both are connected across the poles of the same battery. The voltage drop across each load is equal to ten volts. With this in mind, it is easy to calculate the current through each of the loads:

For load A, using I = E/R, we get:

$$I_A = 10/10 = 1 \text{ A.}$$

For load B, again using I = E/R, we get:

$$I_B = 10/20 = 0.5 \text{ A.}$$

The total current supplied by the battery must be the sum of the current through load A plus the current through load B, or 1.5 A. Unlike the series circuit, where the total resistance of the circuit was obtained simply by adding the resistances of all the series loads, the total resistance of a parallel circuit must be calculated in the following manner:

From Ohm's law, the total resistance R_{total} of the parallel circuit of Fig. 4-1B is:

$$R_{total} = \frac{E_{total}}{I_{total}}$$, but as shown above,

$I_{total} = I_A + I_B$. Therefore,

$$R_{total} = \frac{E_{total}}{I_A + I_B}.$$ But, we know that the current I_A or I_B through a load is equal to the voltage across that load divided by the resistance of the load:

$$I_A = \frac{E_{total}}{R_A} \text{ and } I_B = \frac{E_{total}}{R_B}.$$

If we substitute these expressions for I_A and I_B in $R_{total} = \frac{E_{total}}{I_A + I_B}$ we get:

$$R_{total} = \frac{E_{total}}{\frac{E_{total}}{R_A} + \frac{E_{total}}{R_B}}$$

This complicated looking equation can be simplified by dividing both the numerator and denominator of the fraction by the same quantity, E_{total}. You will recall that this does not change the value of the fraction. Doing this, we get the basic formula for determing the total resistance of a parallel circuit formed by two loads:

$$R_{total} = \frac{1}{\frac{1}{R_A} + \frac{1}{R_B}}$$

(Equation 5-1)

Thanks to modern pocket calculators, any load resistances R_A and R_B are easily plugged into this equation, and the answer appears as fast as the 1/x key can be pressed. It wasn't always that easy. Back in the dark ages of the slide rule, a little trickery had to be used to calculate the total resistance of two loads in parallel, especially since slide rules are useless for addition. It is worthwhile to learn how to perform this calculation without a calculator. The denominator of equation 5-1, $\frac{1}{R_A} + \frac{1}{R_B}$, can be simplified if we multiply the first term by $\frac{R_B}{R_B}$ and the second term by $\frac{R_A}{R_A}$. Since $\frac{R_B}{R_B}$ and $\frac{R_A}{R_A}$ are both equal to one, we haven't changed anything except appearances by doing this, ending up with a fraction denominator that looks like this:

$$\frac{R_B}{R_AR_B} + \frac{R_A}{R_AR_B}.$$

Now, since the denominator terms are both the same, we can add the two fractions, getting: $\frac{R_A + R_B}{R_AR_B}$. This is just another form of denominator of equation 5-1; putting it back in equation 5-1 we get:

$$R_{total} = \frac{1}{\frac{1}{R_A} + \frac{1}{R_B}} + \frac{1}{\frac{R_A + R_B}{R_AR_B}}$$

If we multiply both denominator and numerator of this fraction by R_AR_B, we get a simple, easy to remember formula for the total resistance of two parallel loads:

$$R_{total} = \frac{R_AR_B}{R_A + R_B}$$

(Equation 5-2)

In words, the total resistance of two loads in parallel is equal to the product of the resistances divided by the sum of the resistances, or in short, the product divided by the sum.

In Fig. 5-2, a third load, load C, has been added to the circuit of Fig. 5-1B. Let us assume that it is necessary to calculate the total current I_t in the circuit in order to determine if this third load

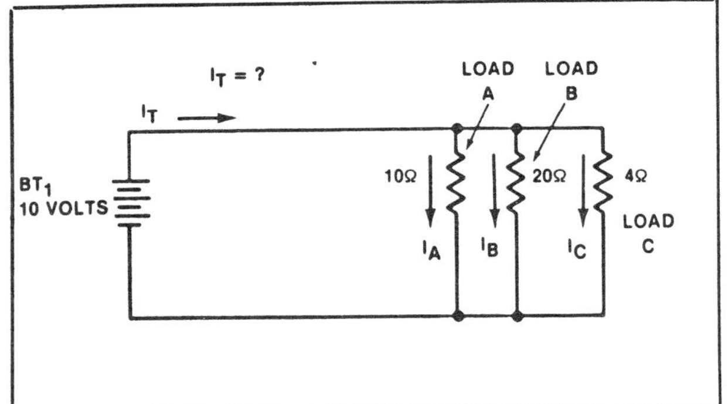

Fig. 5-2 Parallel circuit with three branches.

would cause too great a current flow for our 10-volt battery.

There are three ways in which this might be done. The choice of which to use depends on what else you are required to know and on whether a calculator is available or not. If it is also necessary to know the currents through any of the individual loads, the most direct approach is to use I = E/R and to calculate the individual currents through each resistor:

$$I_A = 10/10 = 1 \text{ A},$$
$$I_B = 10/20 = 0.5 \text{ A},$$
$$I_C = 10/4 = 2.5 \text{ A}.$$

The total current, I_t, is then equal to the sum of the individual currents, or 4 A.

If it is not necessary to determine the individual currents flowing through each load and if a small pocket calculator is available, a modified form of equation 5.1 may be used. For three loads in parallel, the total resistance will be:

$$R_t = \frac{1}{\frac{1}{R_A} + \frac{1}{R_B} + \frac{1}{R_C}}$$

(Equation 5-3A)

In fact, the use of the calculator makes it possible to calculate the total resistance of any number of loads connected in parallel, using equation 5.3B, the general form of equation 5.1,

$$R_t = \frac{1}{\frac{1}{R_1} + \frac{1}{R_2} + \frac{1}{R_3} + \cdots + \frac{1}{R_n}}$$

(Equation 5-3B)

where R_n represents the resistance of the last load in the parallel circuit and the three dots represent all the other terms which have been left out. Before the popularization of pocket calculators, any calculation of this sort for more than three loads with different resistances became the sort of arithmetic battle few people care to face. Now, calculations of any length can be accomplished by pressing just a few keys for each load in the parallel circuit. Having found the total resistance of the parallel circuit, I = E/R may be used to determine the total current required from the battery.

The third method of solving the problem of Fig. 5-2 should be used if only the total current is required and all calculations must be done in the head or on a scrap of paper. This method is a very powerful one in spite of its seeming simplicity, and can be applied to make a simple problem out of a very complex circuit. It is the method of forming an *equivalent circuit*, a circuit containing only one or two components, but having the same properties as a more complex arrangement. Using this method, we can create a circuit with only one load which is equivalent to the three-load circuit of Fig. 5-2. This can be done in the following manner: consider loads A and B, momentarily ignoring load C. Since A and B are in parallel and are connected to the same source of potential difference, we may substitute an equivalent load D, for A and B. Load D, the equivalent load, would have to draw the same current from BT1 as do A and B together. That is, the resistance of load D would have to be the same as the total resistance of the parallel combination of A and B. This is easy to calculate using equation 5-2:

$$R_{A+B} = \frac{R_A R_B}{R_A + R_B} = \frac{200}{30} = 6.67 \text{ ohms}$$

Therefore, it would be possible to replace loads A and B with a 6.67 ohm load without changing the electrical properties of the circuit as far as the battery or load C were concerned. What remains in Fig. 5-3A is a simple two-load parallel circuit. Since we are concerned only with the total current supplied by the battery, we may repeat the process once more, calculating the resistance of a load E which may be substituted for both D and C. the resistance of E will be:

$$R_E = \frac{R_D R_C}{R_D + R_C} = \frac{26.68}{10.67} = 2.5 \text{ ohms.}$$

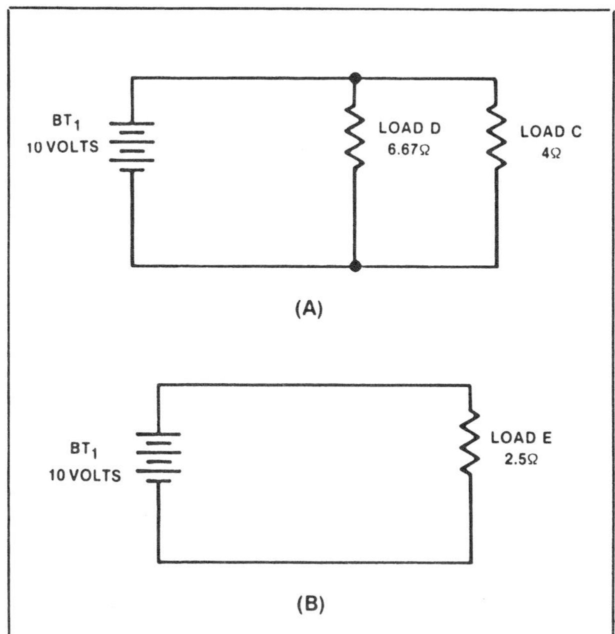

Fig. 5-3 Equivalent circuits — the three load circuit of Fig. 5-2 may be reduced to a two load or a one load equivalent circuit.

Thus, the parallel circuit of three loads in Fig. 5-2 may be replaced by the 2.5 ohm equivalent circuit of Fig. 5-3B. The total current in the equivalent circuit, and therefore the total current supplied by BT_1 of Fig. 5-2, is equal to 10/2.5 = 4 A. (It should be noted that this is the same value for I_t that was obtained through the use of the first method.)

B. Further Contrasts Between Series and Parallel Circuits

The behavior of series and parallel circuits can be demonstrated in a very striking manner by forming contrasting series and parallel circuits with two identical 10-volt batteries and half a dozen 10-ohm loads. (These values have been chosen for ease in computing currents and equivalent resistances; it would not be a good idea to actually try to build these circuits.)

Fig. 5-4A shows a series connection of two 10-ohm loads connected across the 10-volt battery. The total resistance of the circuit is $R_A + R_B$ or 20 ohms and the total current drawn from the battery is $I = E/R$ or 0.5 A. Since this is a series circuit, I_A, the current through load A, is equal to I_B, the current through load B, and to I, the total current drawn from the battery.

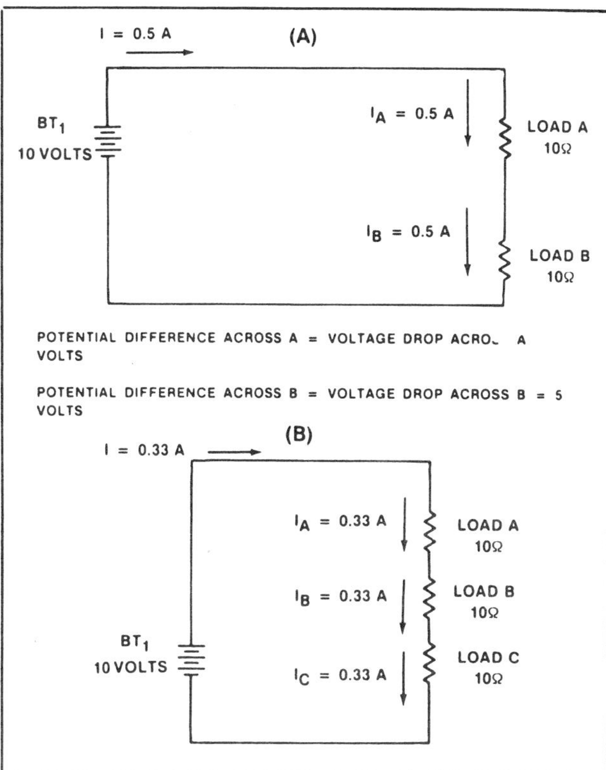

Fig. 5-4 Series circuits.

In a series circuit of two or more loads, the current is the same through all the loads, regardless of the differences in load resistance. It is also equal to the total current drawn from the source. The potential difference across each load in a series circuit depends on the resistance of the load and the current through it ($E = IR_L$, where R_L is the resistance of the load in question). In the circuit of Fig. 5-4A, the potential difference across both load A and load B is the same, 5.0 volts, because the resistances of both loads are equal.

If we add a third 10-ohm load, C, to this circuit as in Fig. 5-4B, the total resistance of the series circuit is *increased* by ten ohms to thirty ohms ($R_A + R_B + R_C$), and the total current through the circuit, which is the same as the current through each of the series loads, is *decreased*. The new total current is I = 10/30 or 0.33 A. The potential difference E = IR across each load also *decreases* to 3.33 volts. Adding another series load then results in an *increase* in the total resistance of the series circuit, and in a *decrease* in the circuit current and in the potential difference across each of the series loads. The potential difference across the entire circuit, however, remains the same, since it is equal to the potential difference supplied by the battery.

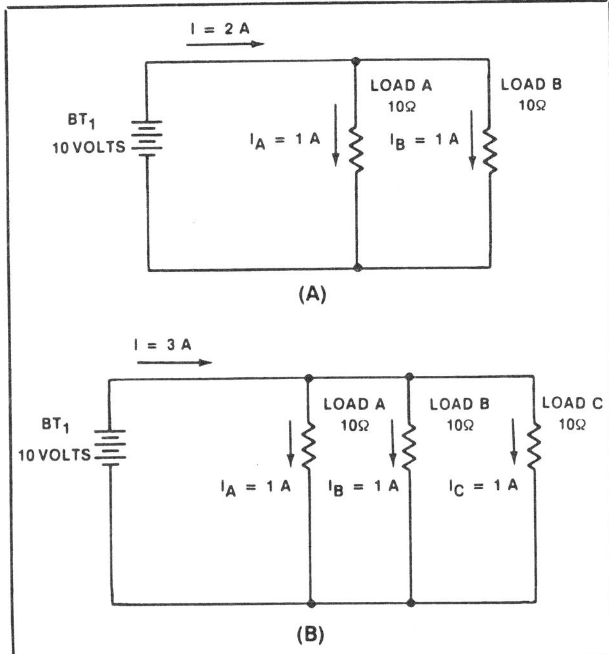

Fig. 5-5 Parallel circuits.

Let us contrast the series circuits described previously with the parallel circuits pictured in Fig. 5-5. The circuit of Fig. 5-5A, like that of Fig. 5-4A, is formed of two 10 ohm loads and a 10-volt battery. The important difference is that the loads are connected in parallel across the battery. The total resistance of the two loads in parallel is:

$$\frac{R_A R_B}{R_A + R_B} \text{ or } \frac{100}{20} = 5 \text{ ohms, or}$$

less than the resistance of either of the loads individually.

In general terms, the combination of two or more loads in parallel produces a resistance that is *less* than the smallest resistance in the combination. The total current supplied by the battery is equal to the potential difference divided by the total resistance of the parallel circuit, or $10/5 = 2$ A, which is exactly four times that found in the series circuit of Fig. 5-4A.

The current through each load in a parallel circuit is not necessarily the same, but in the circuit of Fig. 5-5A, with both load resistances equal to ten ohms:

$$I_A = I_B = \frac{E_{total}}{R_{A \text{ or } B}} = 1 \text{ A.}$$

In other words, exactly one-half the total current flows through each of the two equal loads. The voltage drop across each of the parallel loads is equal to the potential difference across the battery terminals, or 10 volts. In a parallel circuit this would be true regardless of the resistance of the load. It would not be true in a series circuit where the voltage drop across each load is equal to the product of the current and the load resistance.

Addition of a third 10-ohm load C to the circuit of Fig. 5-5A results in the circuit of Figure 5-5B. In this case, the total resistance of the circuit is given by:

$$R_t = \frac{1}{\frac{1}{10} + \frac{1}{10} + \frac{1}{10}} \text{ or 3.3 ohms.}$$

In general terms, addition of another load to a parallel circuit *reduces* the total resistance of the circuit, the total resistance always being smaller than the smallest load resistance in the circuit.

Where the resistances of all the loads are equal, the calculation of the total resistance in a parallel circuit becomes simple; for two equal loads in parallel, the resulting total resistance is half the resistance of either load. Thus, adding a load of equal resistance in parallel to a simple circuit, forming a circuit of the sort shown in Fig. 5-5A, results in a *halving* of the total resistance of the original circuit. Similarly, three equal loads in parallel produce a total resistance equal to one third the resistance of one of the loads, and so on.

With three 10-ohm loads in parallel, as in Fig. 5-5B, the total current I supplied by the battery will be $I = E/R_t = 3$ A. In contrast to the series circuit, addition of another load in parallel *increases* the total current in the circuit. The voltage drop across each of the loads, however, remains equal to the battery potential difference, 10 volts, and the current through each of the loads $I_{A,\ B,\ \text{or}\ C} = E/R_L$ also remains constant at 1 A because neither the battery voltage nor the resistance of an individual load is changed.

C. Series-Parallel Circuits

The majority of the circuits used in avionics equipment are neither purely parallel nor purely series in form. They combine both series and

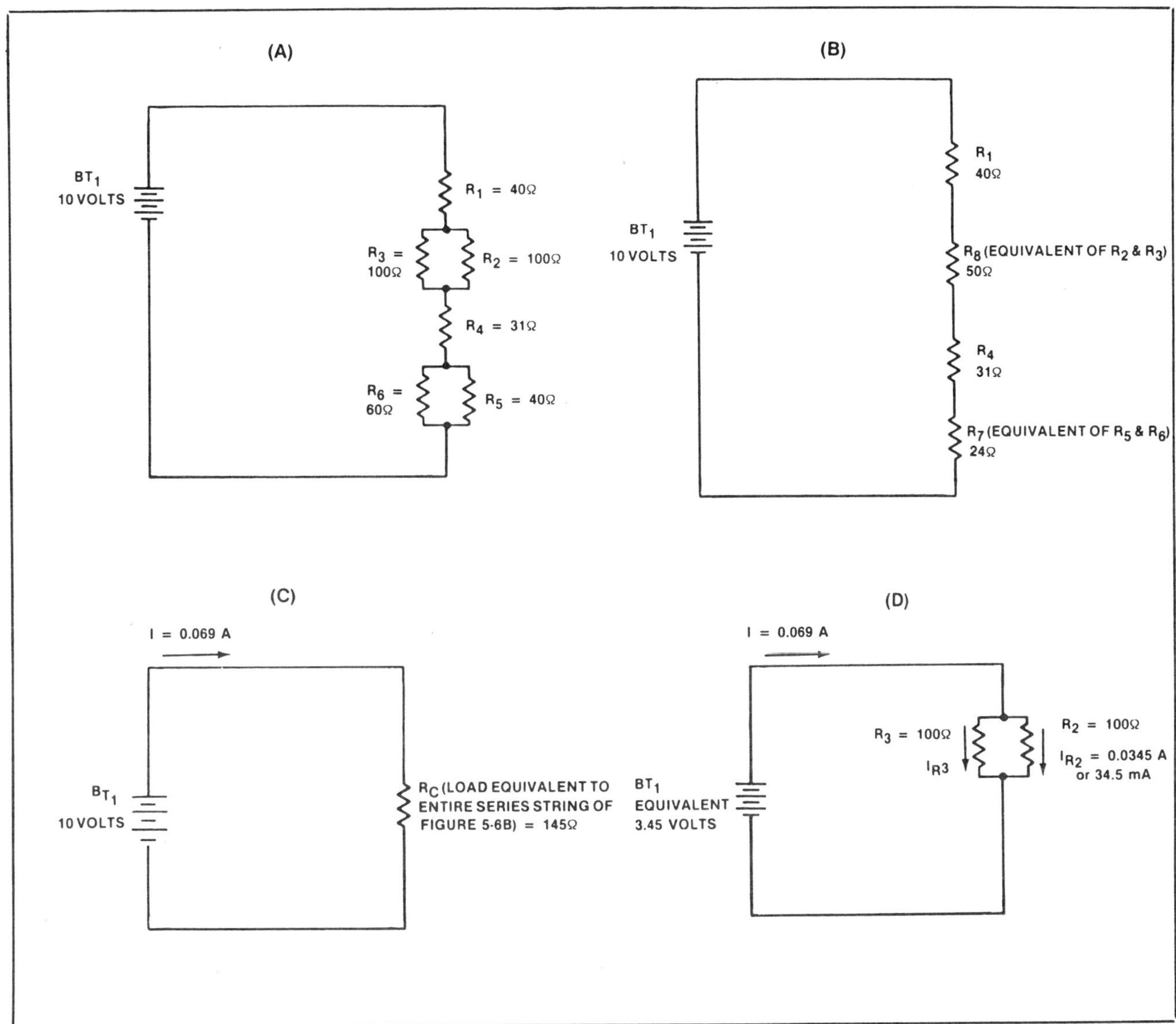

Fig. 5-6 A — Series-parallel circuit. B — Substitution of two equivalent loads. C — Substitution of one equivalent load. D — Circuit of A redrawn to isolate only the section required.

parallel segments to form *complex*, or as they are normally called, *series-parallel* circuits. Fig. 5-6A represents such a circuit.

Using Ohm's law and the method of equivalent circuits, it is possible to calculate the current through and the voltage across any load in a series-parallel circuit. The general method of doing this is to calculate the total circuit current by replacing all the loads in the circuit with a single equivalent load and using Ohm's law. Once this has been done, it is possible to calculate the voltage drops across the individual loads in the circuit and to redraw the circuit so as to isolate the section or sections in which we are interested. Let us apply this method to the circuit of Fig. 5-6A, supposing that it is necessary to know the current through and the potential drop across the load represented by R_2. The total current may be calculated by replacing all six loads with a single equivalent load.

The easiest way to do this is by first reducing the whole circuit to a series string. Using equation 5-2.

$$R_{total} = \frac{R_A R_B}{R_A + R_B}$$

we see that loads R_5 and R_6 may be replaced by R_7, a single 24-ohm equivalent load, and loads R_2 and R_3 by R_8, a single 50-ohm load, yielding the series equivalent circuit shown in Fig. 5-6B. The total resistance of this series string is 145 ohms, as shown in Fig. 5-6C. By using I = E/R we can calculate the total circuit current, 0.069 A or 69 mA. We can now determine the voltage drop across R_8 (the equivalent of R_2 and R_3). It is $E = I \times R = 0.069 \times 50 = 3.45$ volts. We know that the parallel combination of R_2 and R_3, that is, R_8, has a potential difference of 3.45 volts across it.

For this illustrative example we are not interested in the rest of the circuit, so we can redraw it as an equivalent circuit retaining only the parallel combination of R_2 and R_3 with a potential difference of 3.45 volts across it as in Fig. 5-6D. Having reduced the circuit to this point, it is easy to see that by using the methods previously disscussed, the current through R_2 is 0.0345 A, or 34.5 mA, and that the potential difference (equivalent to the voltage drop across it is 3.45 volts.

D. *Power and Electrial Power*

In Sections 2 and 3 we frequently referred to the energy of an electric field or current source and the ability of this energy to do work in moving a charged body such as an electron. Thus, the total energy or work delivered by an energy source such as an electric field or a current source would be the total amount of charge moved times the potential difference against which the charge is moved. This is exactly the same as lifting a weight against the force of gravity, the total energy expended or work done being equal to the total weight lifted times the force of gravity against which the energy is expended. But in most instances, at least in electrical circuits, it is not the total amount of work done that is of prime importance. Instead, it is the rate at which the work is done. This is why the most frequently used measurement of electrical energy is not the coulomb, which is a measure of charge, but the ampere, which is a measure of the charge flowing past a point in one second.

In America today, even though we no longer rely on the horse as a prime mover, the most common unit of power is the horsepower (HP). One horsepower is defined as the amount of power necessary to lift 550 pounds a distance of one foot in one second.

$$1 \text{ HP} = 550 \text{ foot-pounds/second}$$

(Equation 5-5)

The unit of electrical power is the work done in one second when a potential difference of one volt moves an electrical charge of one coulomb through a circuit. This is called one *watt.* Since the movement of one coulomb past a point in one second is an ampere, this definition may be reworded as: one watt is the power expended when a potential difference of one volt moves a current of one ampere through a circuit, or:

$$P \text{ (in watts)} = E \text{ (in volts)} \times I \text{ (in amperes)}$$

(Equation 5-6)

The watt is named for the Englishman James Watt, a late eighteenth-century inventor, considered the father of the modern steam engine. Experiments on relative heating effects have shown that one horsepower is about equal to 746 watts.

$$1 \text{ HP} = 746 \text{ W}$$

(Equation 5-7)

E. *Power Dissipation in Resistive Loads*

Everyone is aware that light bulbs as well as many other household appliances come in various wattages and that the higher the wattage, in general, the more light, heat, etc., produced. Simply stated, wattage is the *power dissipation* of the light bulb or applicance, that is, the amount of electrical power that is expended to heat the glowing wire in the bulb, the electric toaster, or the iron.

In the last section we discussed the mechanism by which heat is produced whenever a current passes through a resistive load, such as a thin wire. But, consider for a moment, the 5-watt automobile bulb pictured in Fig. 5-7. Connected across the 12-volt battery of an automobile, the lamp glows brightly because the thin wire or filament within it is heated by the passage of an electric current. The 5-watt rating printed on the bulb means that it dissipates or uses up 5 watts of power in producing light and heat. From equation 5-6, it is easy to calculate the current through the bulb:

$$P_{watts} = E_{volts} \times I_{amps}$$
$$5 = 12 \text{ I}$$
$$I = .42 \text{ A (rounded off)}$$

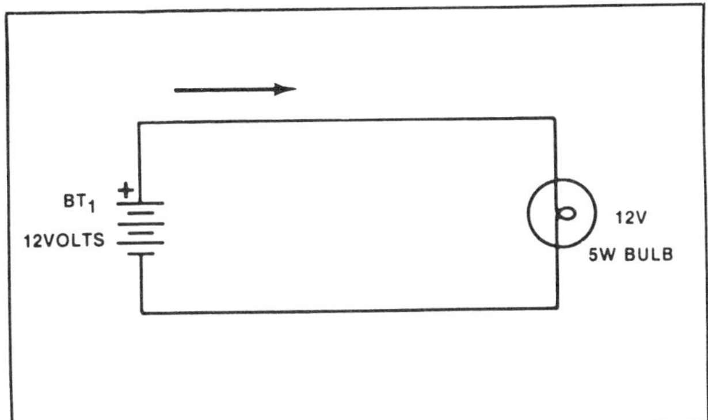

Fig. 5-7 A five-watt bulb connected across a twelve-volt battery.

Given both the current through the lamp and potential difference across it, the resistance may be calculated from Ohm's law: R = E/I = 28.57 Ω.

In reality, the resistance found above is the *hot* resistance of the filament. Its resistance when cold may be a tenth of this or less, and this difference between the hot and cold resistance of indicator bulbs can place a great strain upon a battery or a power supply, even though the cold state lasts for a very short time at switch-on. The effect is more marked when high power appliances such as electric heaters are switched on. This causes a momentary dimming of the electric lights which are connected in parallel with the heater (Fig. 5-8), since the very low resistance of the cold heater element draws excessive current from the line circuit.

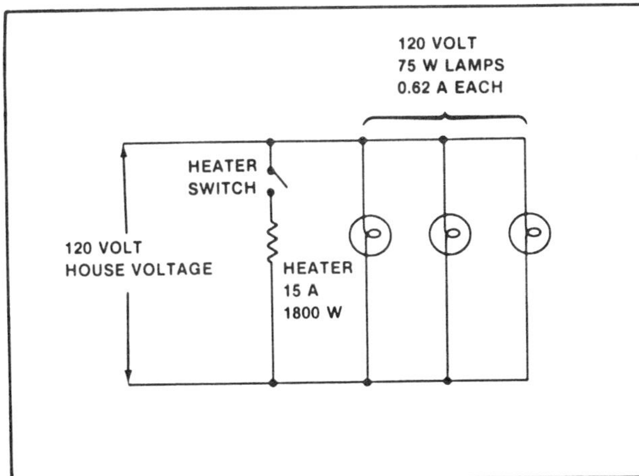

Fig. 5-8 A parallel connection of three bulbs and a high current load across a power line. When the heater is switched on, the lamps will momentarily dim.

What would happen if the 5-watt lamp of Fig. 5-7 were connected across two 12-volt automobile batteries in series as in Fig. 5-9? If you said that it would probably burn out, you are correct. Let us see why.

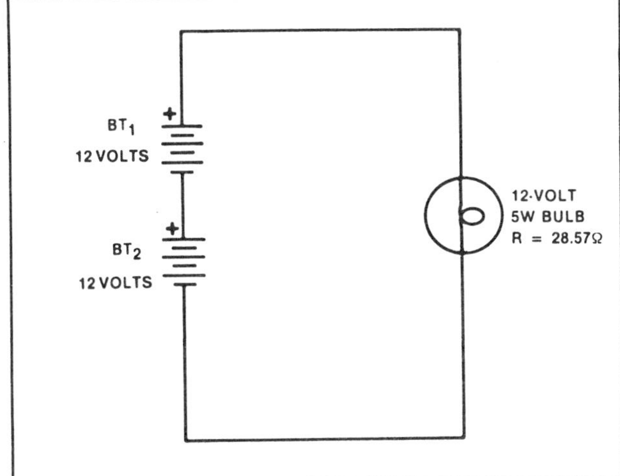

Fig. 5-9 A five-watt lamp connected across too great a potential difference. The lamp will burn out.

Let us assume that the hot resistance of the filament would remain 28.57 ohms as previously calculated (this is actually *not* the case, as we shall see in our discussion of the resistance of wires in Section 6, but we may assume that it is for the moment). What would the power dissipation of the bulb be when connected across two batteries together producing a potential difference of 24 volts? In this simple circuit, by knowing both the voltage across the lamp and the lamp's load resistance, it is simple to find the current using I = E/R. Substitute the current and voltage values in equation 5-6, and derive the power dissipated in the bulb in Fig. 5-9. This is not necessary, because by knowing the potential difference across the load and its resistance, we may determine the power dissipation directly.

We know that:

$$P = E \times I \qquad \text{(Equation 5-6)}$$

and also that I = E/R; substituting E/R for I in equation 5-7, yields:

$$P = \frac{E^2}{R}, \quad \text{or}$$

$$P = \frac{(24)^2}{28.57} = 20.16 \text{ watts.}$$

From this result we see that the 5-watt bulb has to dissipate four times as much power as it was designed to handle. This excess power will probably cause the filament to vaporize, and may even cause the glass portion of the bulb to shatter.

A second form of equation 5-6 is extremely useful when dealing with series circuits made up of resistive loads. This form is obtained by substituting the Ohm's law expression $E = IR$ in equation 5-6, to obtain:

$$P = I^2R \qquad \text{(Equation 5-9)}$$

Since the current through each load in a series string is the same, we can use equation 5-9 to obtain the power dissipation in each of a series string of loads, if the current through the string and the resistance of each load are known.

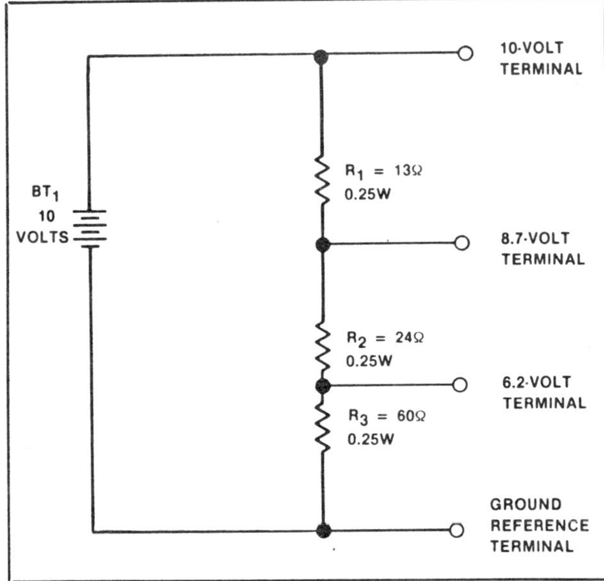

Fig. 5-10 Power dissipation in a voltage divider circuit.

Consider the voltage divider of Fig. 5-10. Would it be possible to construct such a divider using loads rated at 0.25 watt maximum power dissipation? The circuit of this schematic is a practical one, designed to produce potential differences of about 6.2, 8.7, and 10 volts from a single 10-volt battery or other current source. The total resistance of the series string of loads ($R_1 + R_2 + R_3$) is 97 ohms. The circuit current is then $I = E/R = 10/97 = 0.1$ A. Using equation 5-9 we may quickly calculate the power dissipation in each of the loads:

for R_1, $P = I^2R = (0.1)^2 \times 13 = 0.13$ watts.
for R_2, $P = I^2R = (0.1)^2 \times 24 = 0.24$ watts.
for R_3, $P = I^2R = (0.1)^2 \times 60 = 0.6$ watts.

Therefore, the answer to our question would be that it would not be possible to build such a voltage divider using only loads rated at 0.25 w. It would be necessary to use a 1.0 watt load for R_3, and a good idea to substitute a 0.5 watt load for R_2, in order to provide a margin of safety.

A. Self-Test Questions:

Fill in the blanks:

1. If two 80-ohm loads are connected in series, the resulting resistance is ________ ohms.
2. Two 5-ohm loads in parallel may be replaced by an ________ load of ________ ohms.
3. Adding another load in series with two other series-connected loads in a circuit will ________ the circuit current and ________ the voltage drop across each load. The resistance of the entire circuit will ________.
4. Adding another load in parallel with two other parallel loads in a circuit will ________ the circuit current. The resistance of the entire circuit will ________, and the potential drop across each load will ________.
5. In a parallel circuit composed of two loads, the total resistance of the parallel combination R_t = ________ or 1/?.
6. If a lightbulb draws 0.83 A when connected across a 120-volt line, its power rating is ________ watts.
7. If a 150-ohm resistive load is dissipating exactly 0.25 watt, the current through it is ________ mA.
8. A 500-ohm resistive load connected across a 12-volt battery will dissipate ________ watt.
9. A 0.5-watt load may be safely connected across a 120-volt DC line if its resistance is ________ ohms or greater.

10. A 75-watt household lamp draws ________ mA of current from the mains.

B. Problems For Calculation:

1. Calculate the total resistance between terminals A and B in Fig. 5-11. Draw an equivalent circuit using only one load.

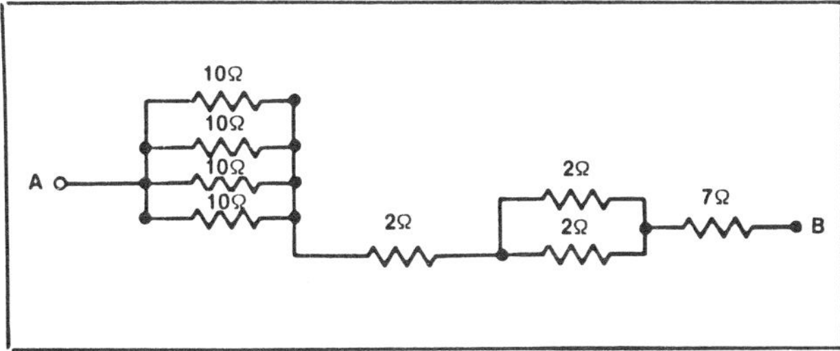

Fig. 5-11

2. Calculate the current through and the voltage drop across each load in Fig. 5-12.

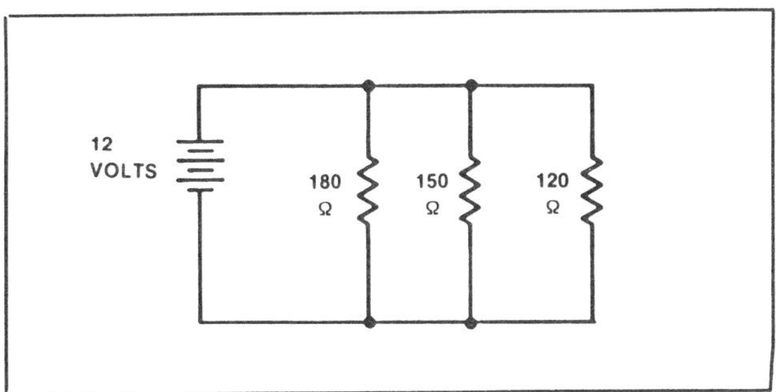

Fig. 5-12

3. What would the resistance of R_X have to be in Fig. 5-13 to reduce the current from the battery to 10 mA?

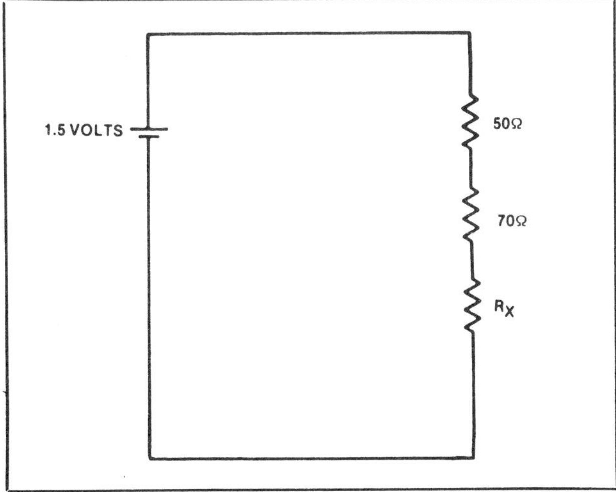

Fig. 5-13

4. Calculate the current through R_X in Fig. 5-14 and the voltage drop across it.

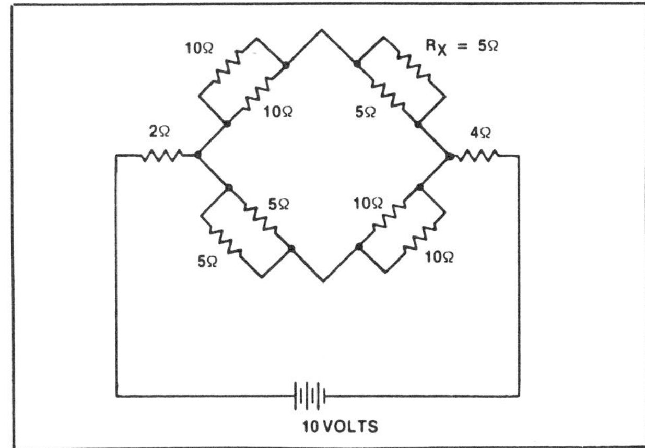

Fig. 5-14

5. Three 100-watt bulbs, a 600-watt toaster, and a 250-watt steam iron are all connected across a 120-volt household line. What is the total current drawn from the line? If the line has a 15 A circuit breaker in it (an automatic switch set to open the circuit if the total current drawn exceeds 15 A) could a 1 kw heater be connected as well without opening the circuit breaker? How much additional power could be drawn from the line without automatically opening the breaker?

C. Experiments and Demonstrations:

1. Make the test set-up shown in Fig. 5-15 and obtain a spool of no. 30 insulated copper wire. Starting with a 25-foot piece of wire, make a variable 2-ohm load by wrapping the wire around a six-inch long, 1/2-inch diameter aluminum or cardboard tube. (Be careful with cardboard, your homemade loads are liable to get hot enough to ignite paper.) Carefully scrape the insulation from the free end of the no. 30 wire and from a strip about 1/4-inch wide along one side of the coil (refer to Fig. 5-16). Place the slider on the load body and tighten the 2-56 bolt so that the dimpled area on the slider contacts the area scraped free of insulation near the end *farthest* from the free end of the coil. Connect the coil free end to terminal A and the slider connecting wire to terminal B of the test set-up. Close the switch and quickly note the ammeter and voltmeter readings. Open the switch (this prevents excessive discharge of the batteries). Calculate the re-

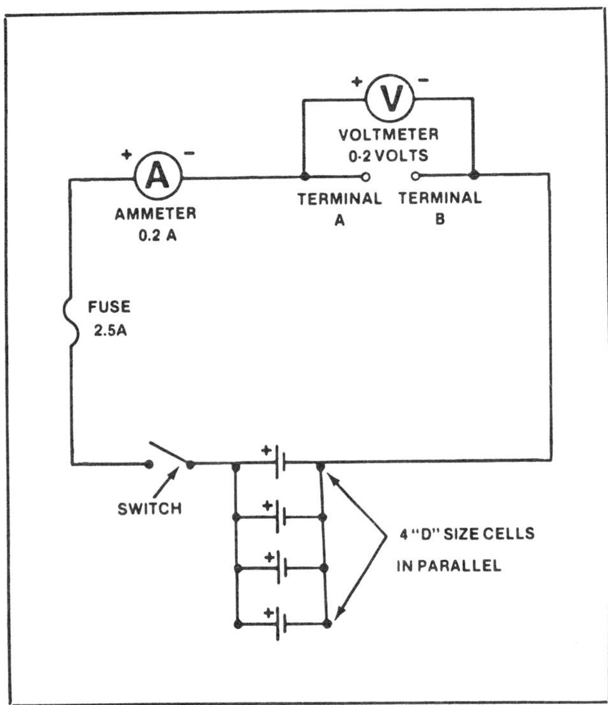

Fig. 5-15

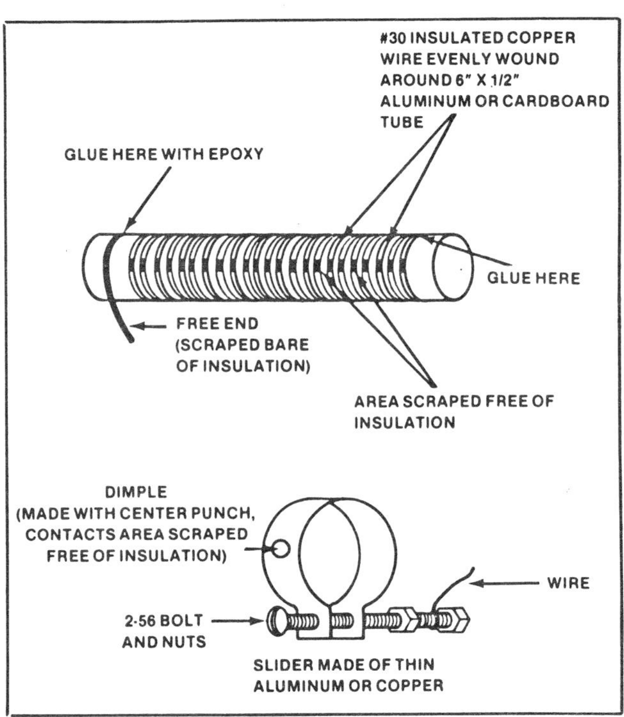

Fig. 5-16

sistance of your load. Note that the resistance may be increased by moving the slider further from the end connected to terminal A or reduced by moving the slider closer to the end connected to A.

a. Set your homemade load as close to 2 ohms as possible.

b. Construct three more 2-ohm loads for use in experiment 2.

c. Calculate the resistance per foot of no. 30 copper wire. (Hint: calculate the resistance of the entire coil and divide by the length of wire used.)

2. Using the test set-up of Fig. 5-15, connect two 2-ohm loads in parallel between terminals A and B. Close the switch and quickly record the ammeter and voltmeter readings. Open the switch.

a. Calculate the resistance of the parallel combination using Ohm's law.

b. Set the slider of a third load about halfway between the ends of the coil. Substitute this equivalent load for the parallel combination of part a. Close the switch and take readings, then open the switch again. Try to adjust your equivalent load so that the readings will be the same as those obtained in part a. What is the resistance of the equivalent load?

c. Repeat part b using the fourth of your loads. Once this is done you will have two one-ohm loads. Connect these in *series* between terminals A and B of the test set-up, close the switch and quickly take ammeter and voltmeter readings. Open the switch. Calculate the resistance of the series loads.

d. Substitute a single 2-ohm equivalent load for the two series-connected 1-ohm loads. Are the current and voltage readings obtained for the 2-ohm equivalent load the same as obtained for two 1-ohm loads in series?

SECTION VI

Introduction to Electronic Components

In the first five sections of this book, we have often referred to some details of components and equipment in passing, but these references have been on an "as needed" basis. In this section and following sections it is intended that the student become familiar with some of the basic components used in avionics, while increasing the range of his theoretical knowledge and skill in the everyday process of solving problems.

In order not to overburden what was meant to be an introductory book, most of the "how-to-do-it" material has been reserved for a later work in this series. Therefore, with the exception of some of the basic information required to perform the experiments and demonstrations found at the end of each section, most of the material presented in the following sections is designed to introduce materials necessary for successful trouble-shooting and maintenance.

The first topic to be covered in this section is a brief introduction to the physical shapes and layouts of avionics equipment, including many construction techniques used. Various features having to do with maintenance are included. Following this is an introduction to the color code, the basic marking and number-encoding system frequently used to mark the resistance of resistors, encode wires, and so forth. Succeeding paragraphs provide detailed discussions of resistivity, and of wires and cables used in avionics equipment.

A. *The Shape of Things*

Even if your experience with electronic equipment is limited to changing transistor radio batteries, you have no doubt noticed how most manufacturers of consumer equipment strain to develop ingenious and very individual shapes, sizes, and construction details for packaging their circuitry. On some miniaturized commercial equipment, the service technician may often spend as much time taking the unit apart and reassembling it as he will put into finding the cause of trouble. Fortunately, most of the electronic equipment and components with which the avionics technician works are standardized with respect to shape, size, and construction details, making it a much easier task to "get into" the actual work of repair.

1. *Ground equipment*

Much of the electronic equipment associated with the "ground side" of guiding, detecting, or communicating with airplanes, regardless of the manufacturer, is designed to permit mounting in standard 19-inch wide electronic equipment racks, sometimes called *relay racks* (Fig. 6-1). The *open rack* consists of two upright steel channels supported on a base. The entire rack may range in height from about 22 inches to about 85 inches depending on where it is to be used. A popular maximum height in vans and trailers is 60 inches and 72 inches is the common maximum height for use aboard ship.

The upright channels are drilled in a standard manner and generally tapped to fit a 10-32 wing bolt, thumbscrew, or other quick release bolt for fastening the equipment to the rack. As shown in Fig. 6-2, units designed for 19-inch rack mounting are equipped with a drilled or slotted front panel to which the component-bearing chassis is attached. Panel width is standardized at 19 inches, while panel heights, depending on mounting hole spacing, may be varied stepwise from a minimum of less than two inches to a usual maximum of about 31-1/2 inches.

Switches, controls, and indicator lamps as well as plug-in units and connections which require frequent changing or removal are mounted on the front panel. The more permanent connections, such as power leads, etc., are made to terminal strips or plug sockets on the rear of the chassis. Since the actual clearance distance between the two upright channels is about 17-3/4 inches, the

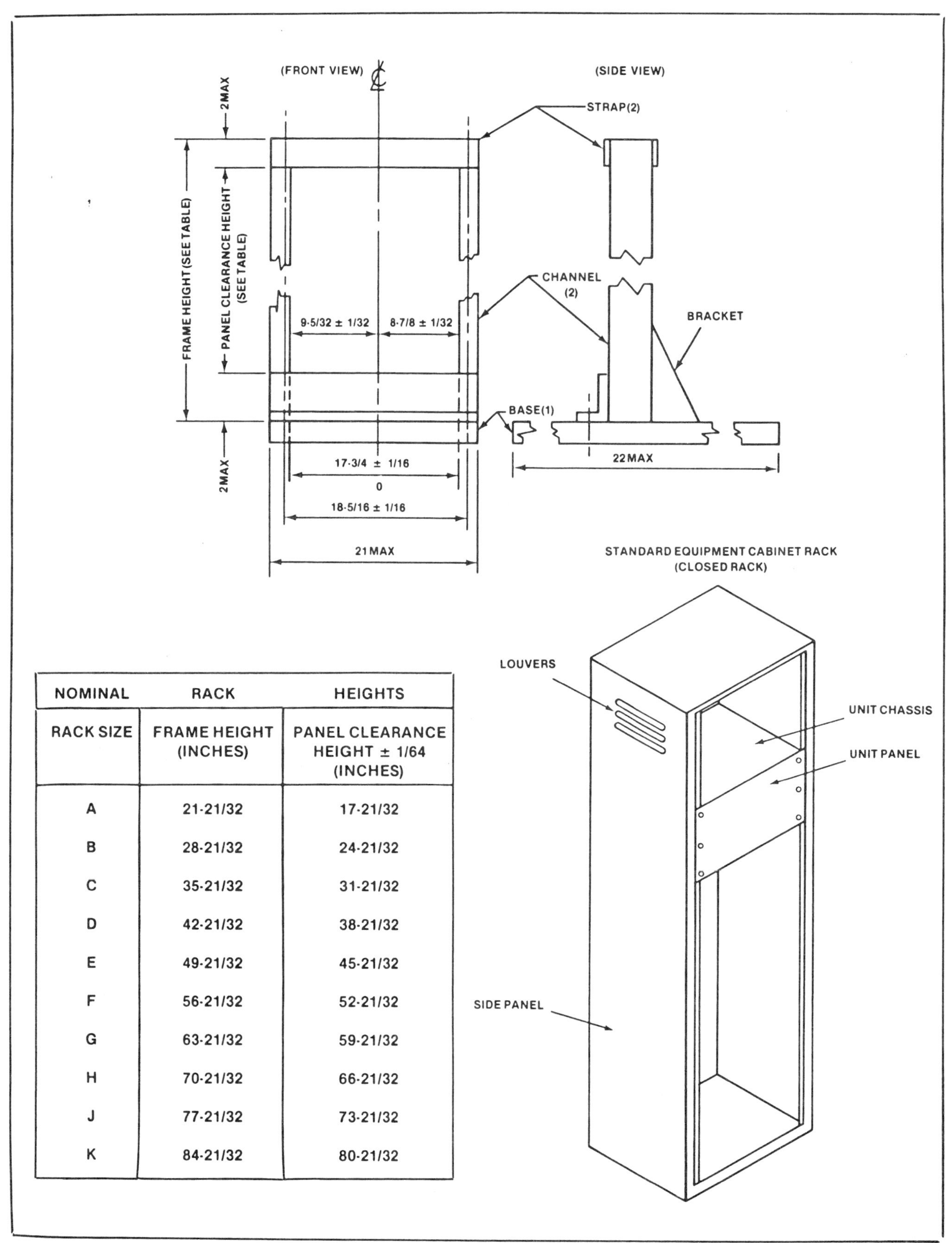

NOMINAL	RACK	HEIGHTS
RACK SIZE	FRAME HEIGHT (INCHES)	PANEL CLEARANCE HEIGHT ± 1/64 (INCHES)
A	21-21/32	17-21/32
B	28-21/32	24-21/32
C	35-21/32	31-21/32
D	42-21/32	38-21/32
E	49-21/32	45-21/32
F	56-21/32	52-21/32
G	63-21/32	59-21/32
H	70-21/32	66-21/32
J	77-21/32	73-21/32
K	84-21/32	80-21/32

Fig. 6-1 Standard 19-inch relay racks used for ground equipment.

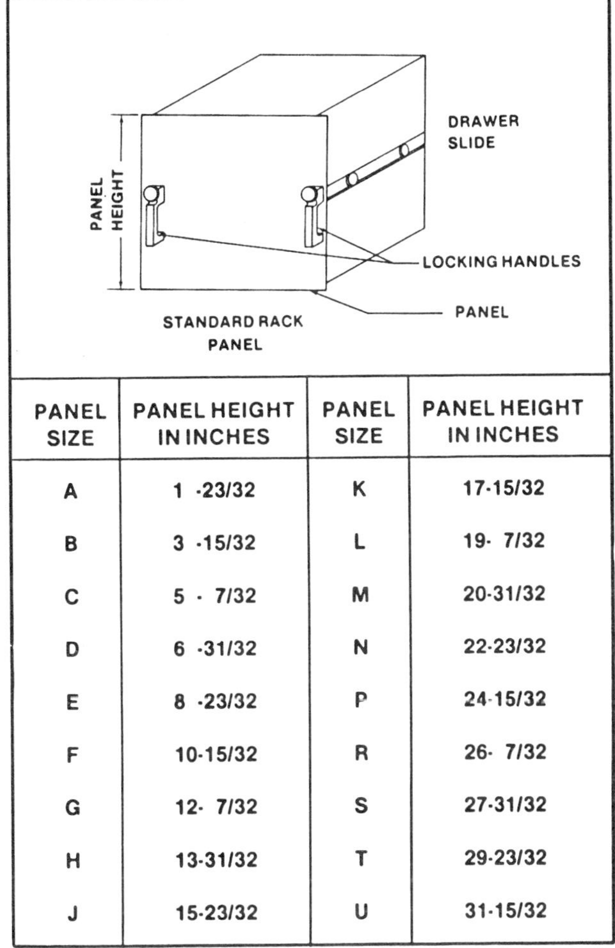

PANEL SIZE	PANEL HEIGHT IN INCHES	PANEL SIZE	PANEL HEIGHT IN INCHES
A	1 -23/32	K	17-15/32
B	3 -15/32	L	19- 7/32
C	5 - 7/32	M	20-31/32
D	6 -31/32	N	22-23/32
E	8 -23/32	P	24-15/32
F	10-15/32	R	26- 7/32
G	12- 7/32	S	27-31/32
H	13-31/32	T	29-23/32
J	15-23/32	U	31-15/32

Fig. 6-2 A sliding drawer mount used for closed standard rack.

actual width of the component chassis must be less. The depth behind the panel is not critical in the open rack, but is generally limited to 17-1/2 inches so that the unit may also be mounted in a closed rack of the sort shown in Fig. 6-1.

The *closed rack*, constructed in a number of ways, provides a convenient means of grouping equipment, perhaps produced by different manufacturers, into a single cabinet system. Instead of rigidly bolting the panel to a pair of upright channels, many modern closed racks use a drawer slider arrangement and locking handles like that shown in Fig. 6-2. This permits the unit to be slid out of the cabinet for repairs.

A useful feature of some drawer arrangements permits the unit to be rotated and locked in place as shown in Fig. 6-3. This makes it especially easy to replace components mounted on the chassis. Connections to the various units mounted in a closed rack are made via multi-wire cables to terminal boards or plugs at the rear of each unit chassis.

2. *Airborne equipment*

A great premium is placed on compactness and light weight in airborne electronics. Because of their fixed width, the 19-inch wide panels of the relay rack will not make the best use of the areas which may be set aside in an airplane for elec-

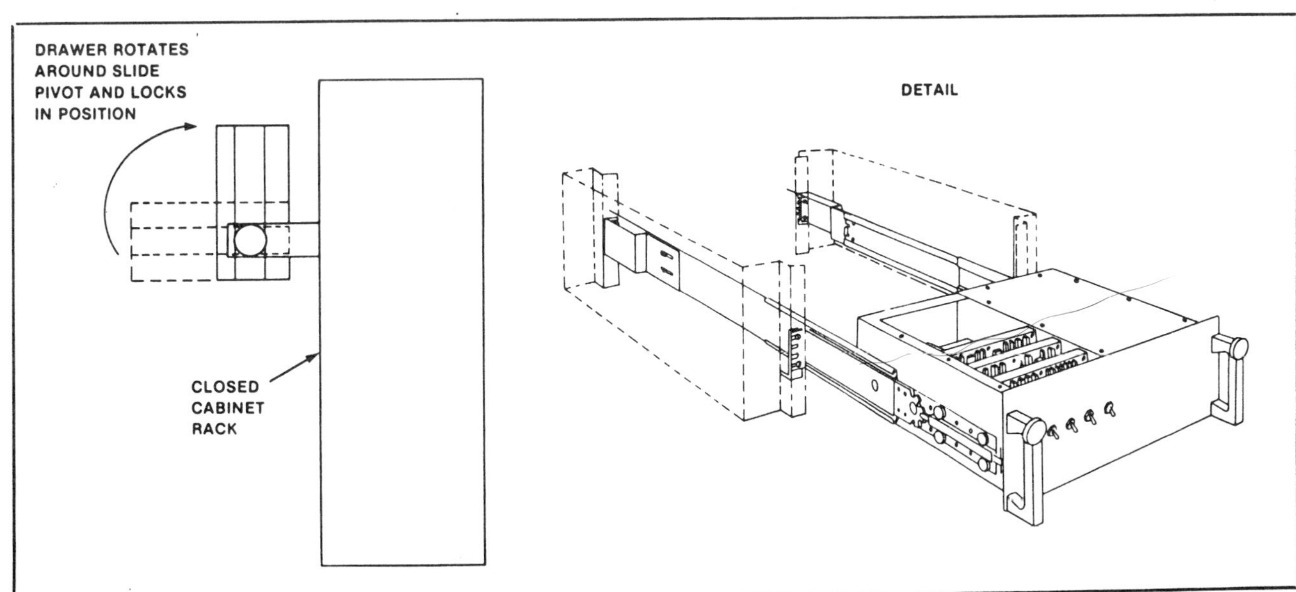

Courtesy of Wilcox Electric

Fig. 6-3 Rotating drawer equipment mounting. The entire unit may be slid out of the rack and then rotated for easy access to components.

tronic equipment. On the other hand, the best solution to the problem, from the point of view of utilization of available space, is to design electronic units to exactly fit the space available in each model of airplane. This, however, would probably be met with great opposition by commercial equipment manufacturers, as a different shaped case would be required for each model of aircraft in which the equipment was used. For exceptional projects, such as certain equipment for use on military aircraft, this is necessary.

The first and still most widely accepted attempt at a standardization of airborne electronics equipment size and installation appeared in 1940, when Aeronautical Radio, Inc. (Arinc), an association of American airline operators, decided to standardize the radio installation on the DC-4 airplane then in use on the routes of some of the larger companies.

The product of this decision, Arinc specification 404, eventually became an industry standard which determined the shape and mounting details of much airborne equipment. Specification 404 called for the standardization of electronic equipment to fit the ATR racking system, based on a unit called the ¼ ATR (Fig. 6-4).

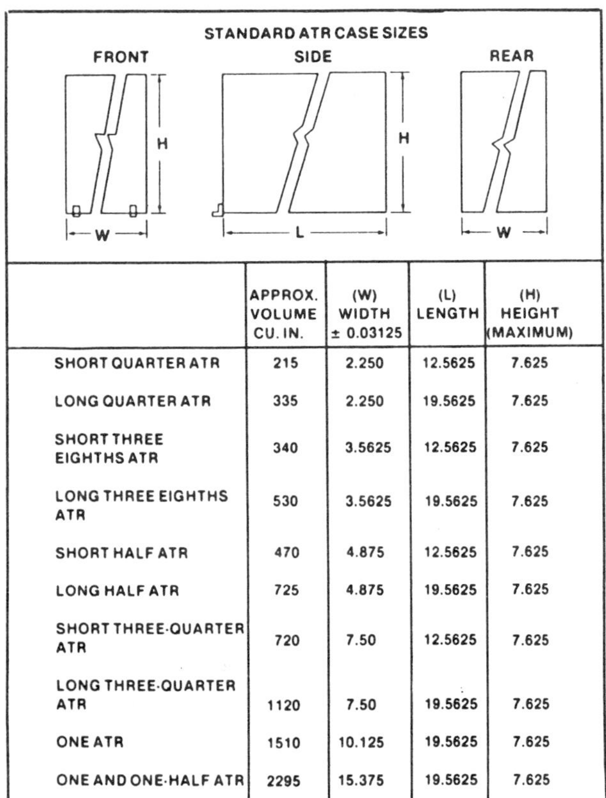

	APPROX. VOLUME CU. IN.	(W) WIDTH ± 0.03125	(L) LENGTH	(H) HEIGHT (MAXIMUM)
SHORT QUARTER ATR	215	2.250	12.5625	7.625
LONG QUARTER ATR	335	2.250	19.5625	7.625
SHORT THREE EIGHTHS ATR	340	3.5625	12.5625	7.625
LONG THREE EIGHTHS ATR	530	3.5625	19.5625	7.625
SHORT HALF ATR	470	4.875	12.5625	7.625
LONG HALF ATR	725	4.875	19.5625	7.625
SHORT THREE-QUARTER ATR	720	7.50	12.5625	7.625
LONG THREE-QUARTER ATR	1120	7.50	19.5625	7.625
ONE ATR	1510	10.125	19.5625	7.625
ONE AND ONE-HALF ATR	2295	15.375	19.5625	7.625

Fig. 6-4 Standard ATR case sizes.

As shown in Fig. 6-4, the maximum height of all units was originally fixed at 7.6 inches, the depth at either 19.5 inches, or, for a shortened version at 12.5 inches. The width could be varied in multiples of the basic ¼ ATR width of 2-1/4 inches. Later, provision was made for very small modules which could be stacked in a ¼ ATR frame.

The ATR system calls for connections to be made at the rear of each unit, by means of a connector which also serves to hold the unit in place. A special mounting lip on the front of each unit is engaged by quick-release clamps or wing nuts, enabling a unit to be removed and another to be installed in its place in a matter of seconds.

B. Construction Details

Almost all electronic assemblies, whether ground or airborne, are built around an aluminum or plated steel body, or *chassis* as it is normally called. The shape of the chassis is dictated by the racking system, if any, and the size and type of components to be mounted on it. Before the popularization of printed circuit boards, the electronic components were mounted directly on the chassis. They were interconnected by wires that were integral parts of the components themselves, as in the case of capacitors and resistors, or by means of insulated hook-up wires. Such wiring is now called *point-to-point wiring* and is still frequently found. For purposes of shockproofing the units wired with point-to-point techniques, particularly when heavy components are involved, various types of terminals and tie points anchored to the chassis are used (Fig.6-5).

Although adequate in the early days of avionics and electronics in general, point-to-point wiring has fallen into disfavor in recent years for a number of reasons: it involves a great deal of expensive hand labor, and provides an entrance for human error. Further, it requires greater space than do more modern techniques, and is out of keeping with the requirements of miniaturization so essential in airborne equipment. Finally, a point-to-point wired chassis, especially in complex equipment, is difficult to troubleshoot and repair. Yet, in spite of these faults, it has not proved possible to eliminate the point-to-point

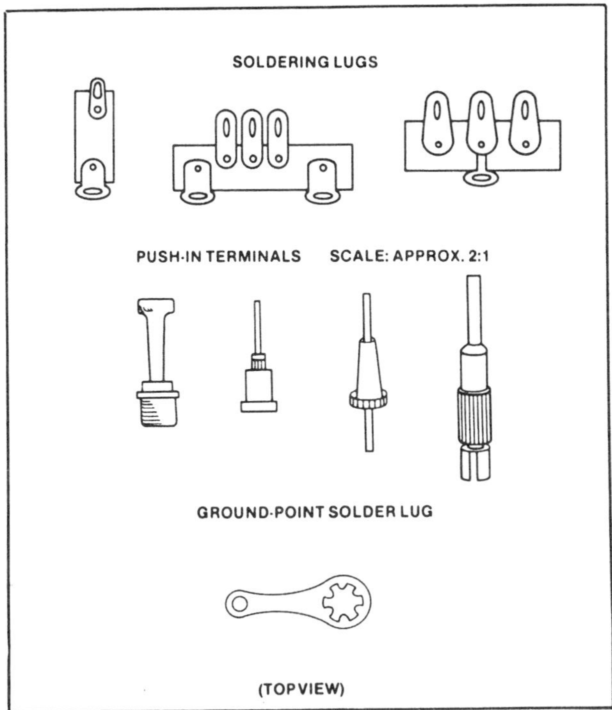

Fig. 6-5 A variety of tie-points and soldering lugs used in point-to-point wiring.

method. This type of construction is still found in new equipment today, usually along with newer methods.

The first of the replacements for point-to-point wiring to be tried was that of modularization of complex pieces of equipment. The idea here was to divide the unit into a number of functional modules which plugged into sockets in the main chassis. In many cases, the modules themselves were not considered repairable and were made up of components embedded in a solid encapsulating medium such as tar or plastic. This construction technique followed the invention of the transistor, which had a much longer average lifetime than the tube and which produced much less heat.

Repair of such modularized units consisted of finding the faulty module and replacing it. This was an expensive and seemingly wasteful approach, but it was economically justified when the out-of-service time of an expensive piece of equipment could be drastically reduced.

The most important recent breakthrough in construction techniques came with the perfection of the *printed circuit board* (PCB) and the *integrated circuit*. The printed circuit board enabled the replacement of individually routed hook-up wires by copper strips bonded to a glass fiber or plastic board. Small components could then be automatically installed on the boards and joined to the copper strip by mass production techniques. This, coupled with the integrated circuit technology that makes it possible to reduce highly complex circuits to small, uniformly sized packages, contributed much to change the type of work performed by the avionics technician.

In most modern equipment, the chassis has become a simple carrier of connectors into which the printed circuit boards are plugged. First-line troubleshooting now involves determining the board on which the fault has occurred and replacing it with a board known to be good. The faulty board may then be repaired at leisure, without putting the entire unit out of service.

Plug-in printed circuit boards thus combine the ease in first-line troubleshooting of the plug-in modules with the ability to isolate and replace a single faulty component as in the point-to-point system.

C. *The Color Code*

One of the most obvious things about a complex electronic assembly with its various covers removed, are the bright colors that decorate various components and hook-up wires. Since most electronic components are rather small, it is difficult to print much information such as resistance, capacitance, tolerance, and voltage value on the component itself. Even in those cases where the component value may be printed on the component, it seems that the most important piece of information is always placed so as to be unreadable when the component is connected in a circuit. By using patterns of color, with each color representing a number, a surprising amount of information may be encoded on a component only half an inch long, or on a wire that would be too thin to bear any printed information.

1. *Functional color coding of wires*

One way of finding a path through a jungle of point-to-point wiring is the method of using particular colors to distinguish wires by their function. Initially designed for vacuum tube circuits, the system was later adapted to transistors as well (Table 6-1). Although it never achieved complete acceptance, it is widely used, particularly in military equipment.

CIRCUIT FUNCTION COLOR CODES		
Color	Vacuum Tube Circuit	Transistor Circuit
Black	Grounds, grounded tube elements, and return lines	Grounds
Brown	Heaters or filaments	......
Red	Positive high voltage supply line	B+ supply line
Orange	Screen grid lines	......
Yellow	Cathodes	Emitter circuits
Green	Control grid circuits	Base circuits
Blue	Plate circuits	Collector circuits
Violet	Negative high voltage line	......
Gray	A/C power lines	......
White	Miscellaneous non-ground lines	......

Table 6-1

Many of the terms in Table 6-1 may be strange to the reader at present, but it is worthwhile to become familiar with these standard function color codes so that experimental hook-ups given in this text can be made using the appropriate colored wire.

2. Numerical color coding of wires

Another way to solve the problem of identifying a particular wire in a complex point-to-point wired unit or in a cable formed of a number of wires laced together in a bundle (commonly called a *wire harness*), is to assign a number to each wire. This number may be marked on the schematic diagram or referred to in a wiring list or wiring diagram, and actually recorded as a part of the wire by using plastic insulation of an appropriate color (Table 6-2).

If more than ten, but fewer than one hundred wires must be identified, two-color insulation

NUMBER COLOR CODES	
Number	Color
0	black
1	brown
2	red
3	orange
4	yellow
5	green
6	blue
7	violet
8	gray
9	white

Table 6-2

may be used, the background or predominant color representing the tens digit and a narrow spiral stripe encoding the units. Wire number fifty-three, therefore, would have a green background with a thin orange stripe, wire number seven would have a black background (0 as the tens digit) and a thin violet stripe. Wire number sixty-six would be a solid blue (blue background and blue stripe). It should be noted that although the system described above is quite a common one, it is by no means universally used.

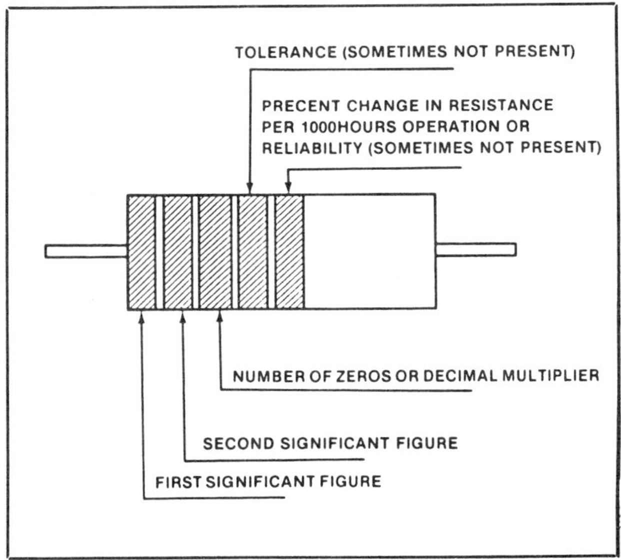

Fig. 6-6 Resistor color banding. Three to five color bands may be used.

RESISTOR COLOR CODING					
Color	Represents				
	band 1	band 2	band 3	band 4	band 5
black	not used	0	x 1 (10^0)	not used	
Brown	1	1	x 10 (10^1)	1%	1%
Red	2	2	x 100 (10^2)	2%	0.1%
Orange	3	3	x 1000 (10^3)	3%	0.01%
Yellow	4	4	x 10,000 (10^4)	4%	0.001%
Green	5	5	x 100,000 (10^5)	not used	not used
Blue	6	6	x 1,000,000 (10^6)	not used	not used
Violet	7	7	not used	not used	not used
Gray	8	8	not used	not used	not used
White	9	9	not used	not used	not used
Gold	not used	not used	x 0.1 (10^{-1})	5%	not used
Silver	not used	not used	x 0.01 (10^{-2})	10%	not used
No color band	not used	not used	not used	20%	not used

TABLE 6-3

3. Color coding of resistor values

The use of color bands to encode the value of resistors has become almost universal for most types of resistors. As shown in Fig. 6-6, the color banding consists of from three to five bands which give the resistance value in ohms, the tolerance of the resistor (the percentage by which the actual resistance of any particular resistor may depart from the marked value) and sometimes, the percentage change of resistance that is to be expected per thousand hours of operation, or the reliability of the resistor.

As shown in Table 6-3, the first two color bands represent the first two digits of the resistance, the third band the decimal multiplier (the number of zeros following the first two digits). The fourth band is the tolerance within which the actual resistance may depart from the marked value.

The fifth band, if present, gives the percentage change in value per thousand hours of operation, or the expected failure rate per thousand hours of operation.

When you first attempt to read the color code, it is sometimes difficult to tell which is the first color band, especially on the smaller sized resistors, if all five bands are present. One way of getting around this problem is to look for the tolerance band, which is the fourth color band. Resistors used in avionics equipment usually have a tolerance rating of 5% or 2%, so that either a gold band, or less often, a red band will appear in this position. More rarely, tolerances of 1% or 10% are used, shown by a brown or a silver band respectively. Once the tolerance band has been spotted, it is easy to determine the order of the bands.

A resistor marked brown, black, red, gold, for example, has a nominal resistance of 1000 ohms or 1 k ohm, although its actual measured resistance may be as high as 1050 ohms or as low as 950 ohms. This is a small variation and has little or no effect when a resistor with the highest possible tolerance value (1050Ω) is substituted for one with the lowest possible tolerance value (950Ω) in a circuit. In cases where this variation is important, 2% or even 1% tolerance resistors must be used.

As another example, consider a resistor with the following color bands: white, brown, gold, gold, yellow. Since a gold band cannot be in the second position, the white band, representing the number 9, must be the first significant figure. The second significant figure is the number 1 represented by the brown band. The multiplier, the third band from the beginning, is gold, representing a value of 0.1. The nominal resistance of this resistor is therefore 9.1 ohms. The tolerance is 5%, as given by the fourth band, so that the initial resistance may be as high as 9.55 ohms or as low as 8.65 ohms. The variation in resistance per 1000 hours of operation is encoded by the final yellow band and is .001 per cent per thousand hours.

There seem to be two glaring deficiencies in this system. First, it is not possible to encode more than two significant figures, so that it would not be possible to show the resistance of a particular resistor as 156,000 ohms or 15.6 K ohms. This would require three significant figures. Second, the wattage rating of the resistor seems to be ignored. As we shall see when we discuss resistors in detail in Section 7, neither of these situations is really a problem.

Before the almost universal adoption of the color banding system, the so-called body-end-dot system of encoding resistor values (Fig. 6-7) was widely used. Although it is highly unlikely that anyone will ever see a resistor of this sort outside a museum, the system has been included for the sake of completeness.

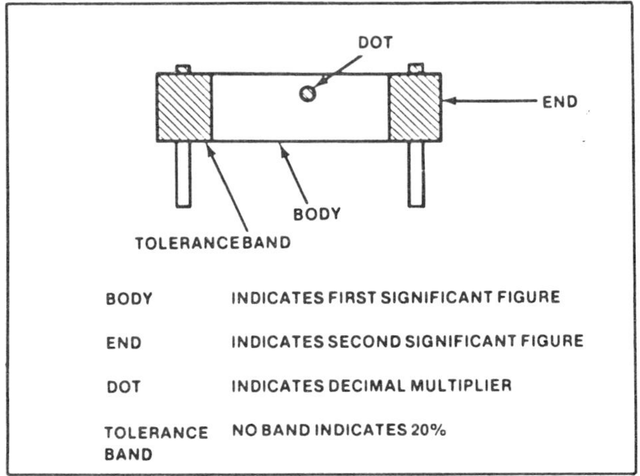

Fig. 6-7 Body-end-dot color code. This system is seldom seen today.

4. Color coding of capacitor values

The use of color coding on capacitors was neither as universal nor as standardized as it has been on resistors. Today, most of the capacitors used in avionics equipment have their values stamped or lettered on them, or, more rarely, bear a serial number that may be used to reference a manufacturer's catalog or military specification.

The most uniformly used color coding system for capacitors was that used for tubular ceramic capacitors (Fig. 6-8). Except for the fact that the tolerance for capacitors of 10 pf or less is given in pf rather than in percent, the color coding of bands two through five closely resembles that used for bands one through four of resistors. The greatest difference is in the first and wider band which gives the rate of change of capacitance per degree centrigrade.

Although the changes shown in the color code table of Fig. 6-8 seem very small, 1 part per million being only 0.0001 per cent, the substitution of a −470 ppm coefficient capacitor for a zero coefficient component or vice versa could result

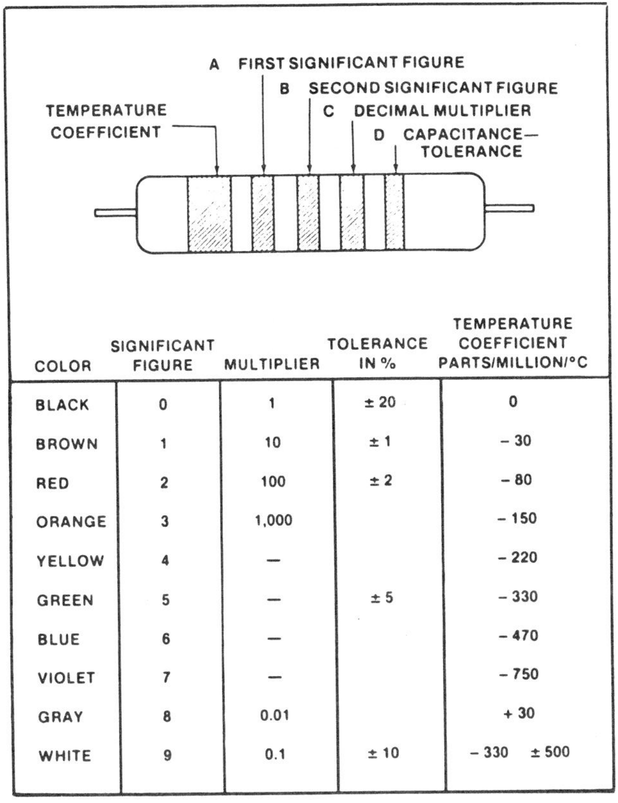

COLOR	SIGNIFICANT FIGURE	MULTIPLIER	TOLERANCE IN %	TEMPERATURE COEFFICIENT PARTS/MILLION/°C
BLACK	0	1	± 20	0
BROWN	1	10	± 1	− 30
RED	2	100	± 2	− 80
ORANGE	3	1,000		− 150
YELLOW	4	—		− 220
GREEN	5	—	± 5	− 330
BLUE	6	—		− 470
VIOLET	7	—		− 750
GRAY	8	0.01		+ 30
WHITE	9	0.1	± 10	− 330 ± 500

Fig. 6-8 Tubular capacitor color code. Like the body-end-dot resistor code, this system is rarely seen today.

in serious detuning of a circuit as a unit warmed up. This is especially true since the temperature coefficient of certain capacitors is specially chosen to cancel the effects of temperature on other components. Rarely, color dots are substituted for color bands two through five.

The color coding illustrated in Figs. 6-9 through 6-11 represents a number of different systems

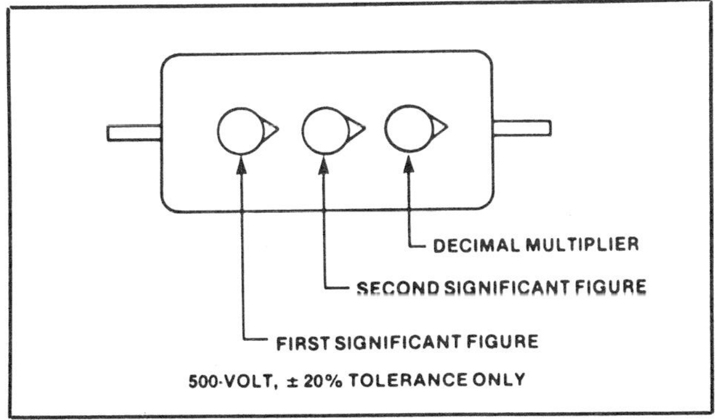

Fig. 6-9 The three-dot capacitor color code system was common during WWII.

used primarily for capacitors made of mica and encased in plastic. These codes are not often used today and have been included mainly for the sake of completeness.

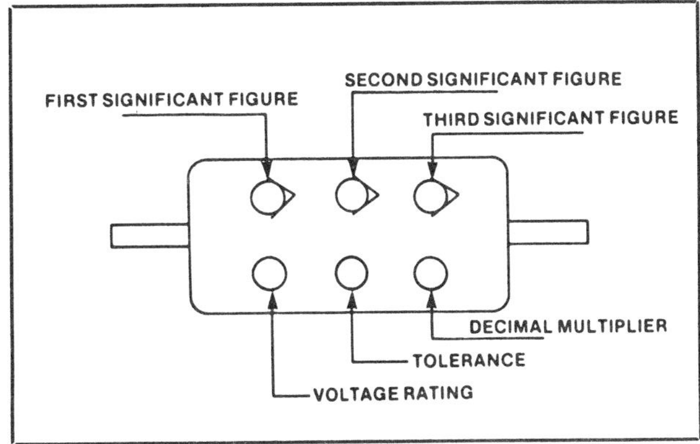

Fig. 6-10 An obsolete system, the six-dot capacitor color code, is seldom seen today.

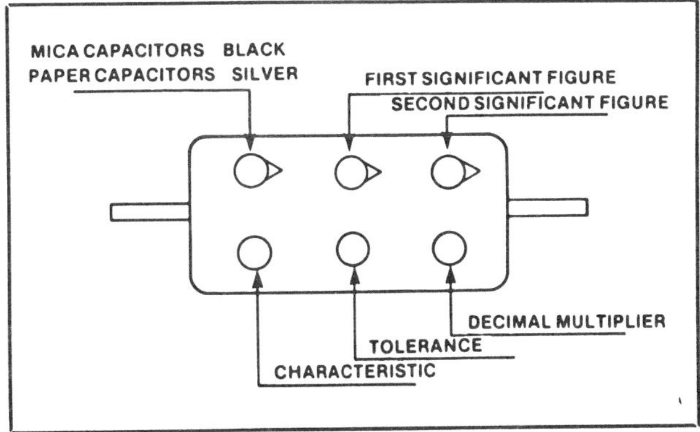

Fig. 6-11 The system most commonly used during WWII was the six-dot military color code.

5. *Color coding of inductor values*

Since the introduction of ferrite cores made of powdered iron embedded in an epoxy or ceramic base, the size of inductors could be held more constant than was possible in the days when an inductor consisted of a coil of wire wound on a waxed cardboard cylinder. Because of the uniformity in size, the advantages of a color coding system became clear, and the system shown in Fig. 6-12 is becoming more common. Notice that here, unlike in the resistor system, the multiplier band is sometimes situated between the first and second figure bands. For inductors greater than 1 mH, the order of the color bands is usually military specification silver band, first significant figure, second significant figure, multiplier, and tolerance band.

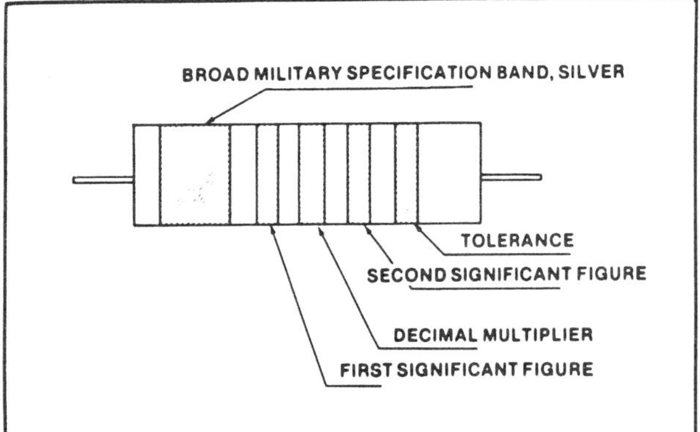

Fig. 6-12 Presently coming into wide use in avionics is the inductor color code.

D. Conductors

As mentioned in Section 1, the purpose of avionics equipment is to direct and control energy to do specific tasks. Electrical energy, as we have seen, is usually moved from place to place by means of plastic-coated conductive wires made of copper, or copper strips bonded to an insulating board.

Although silver, gold, and to a lesser extent aluminum wires are used in electronics, copper is by far the most used material, and every year vast quantities of copper are processed for use in electronic equipment. The reason copper is so much preferred over all the cheaper metals (gold and silver being much too costly for use where their particular properties are not especially required) becomes clear when the resistivity of the various materials is considered.

1. Resistivity of metals

In our discussion of resistance, it was shown that the opposition to the flow of an electric current is an atomic and crystal structure phenomenon and is related to the temperature of the conductor. More exactly, we may say that the actual DC resistance of a piece of metal formed into a wire depends upon four things:

a. the material of the wire,
b. the length of the wire,
c. the thickness of the wire
d. the temperature of the wire.

If wires of equal length and thickness are formed of several different metals, it will be found that the resistance of these wires varies greatly. If we make a proportion of this sort:

$$\frac{\text{Resistance of wire of material X}}{\text{Resistance of heat-treated copper wire}}$$

we may produce a table like that of Table 6-4 showing the relative resistivity of some common metals.

METAL	RELATIVE RESISTIVITY WHEN COMPARED TO COPPER
Aluminum	1.6
Brass (common)	4.2
Cadmium	4.4
Copper	1.0
Gold	1.4
Iron	5.6
Lead	12.7
Nichrome	58.0
Silver	0.9
Steel	10.5
Tin	6.5

TABLE 6-4

The table clearly shows that the best conductor among the metals listed is silver. A length of silver wire will possess a resistance only nine-tenths as great as an equal length of copper wire of the same thicknes. So, if we are interested in reducing the resistance of a circuit to an absolute minimum, it might prove advantageous to use silver instead of copper, but because of the high cost of silver with respect to copper, this is rarely done in DC circuits. Although for certain special applications, silver or silver plated copper are necessary, regardless of the increased cost.

As shown in Table 6-4, gold and aluminum, in addition to silver and copper, also possess low relative resistivity and are therefore good conductors. Each of these metals possesses properties that are useful in electronics. Copper is an excellent conductor, relatively inexpensive and easy to work with, so it is the most commonly used. Gold or gold plating resists corrosion and oxidation and is particularly useful for switch contacts and sliding contacts, although its cost is the highest of the four.

At normal temperatures, silver is the best conductor commonly known and although it will tarnish (form a black silver oxide layer when exposed to the air), its performance as a conductor is not seriously affected. This is because the resistivity of the silver oxide tarnish is almost as low as that of gold. Being considerably cheaper than gold, silver or silver plating is used extensively in switches, contacts, and especially in radio circuits when the lowest possible circuit resistance is required.

Aluminum, the poorest of these four conductive metals, is by far the lightest, which makes it highly useful for airborne equipment. Until recently, however, it was very difficult to fashion good electrical connections with aluminum wires. With the invention of ultrasonic soldering techniques, though, the widespread use of aluminum wire to save weight, particularly in motors and transformers, is on the verge of becoming a reality.

Low resistance is not always what is needed. In the experiments at the end of the last section, we needed about twenty feet of fine number 30 wire to make a two-ohm resistor. If a nichrome wire of equal thickness had been used, a length of less than five inches would have been sufficient.

In the mks system used in this book, the resistivity of metals is given in ohm-meters and is symbolized by the Greek letter ρ (rho). It is the actual resistance measured between opposite faces of a one meter cube of the metal.

Material	Resistivity (ohm-meters)
Aluminum	2.63×10^{-8}
Carbon	3.5×10^{-5}
Copper	1.72×10^{-8}
Glass	10^{10} to 10^{14}
Iron	10×10^{-8}
Lead	22×10^{-8}
Mercury	9.4×10^{-7}
Mica	10^{11} to 10^{15}
Nichrome	1.0×10^{-6}
Silver	1.47×10^{-8}
Wood	10^{8} to 10^{11}

Table 6-5

Table 6-5 gives the resistivity in ohm-meters of some common materials. Notice that the difference between the metallic conductors and the insulators such as glass and mica is very great, on the order of 10^{18} times!

As mentioned earlier in this section, the actual resistance of a wire depends on the resistivity of the material, its length, and its thickness. Momentarily disregarding the effect of temperature, this can be expressed by the equation:

$$R = \frac{\rho L}{A}$$

(Equation 6-1)

where R is the resistance in ohms, ρ the resistivity of the material in ohm-meters, L the length in meters, and A the area of a cross section of the material in square meters. Since the cross section of a round wire is a circle, the area A of the cross section will be $0.25\pi d^2$, where d^2 is the square of the wire diameter in meters, and π is 3.14. Equation 6-1 can then be rewritten as:

$$R = \frac{\rho L}{0.25\pi d^2}$$

(Equation 6-2)

Example calculation: Calculate the resistance of 300 meters of American Wire Gage no. 30 copper wire (diameter = 0.00025 meter).

Solution: From table 6-4, the resistivity of copper is 1.72×10^{-8} ohm-meters. Using equation 6-2 we get:

$$R = \frac{\rho L}{0.25\pi d^2} =$$

$$\frac{1.72 \times 10^{-8} \times 300}{0.25 \times 3.14 \times (2.5 \times 10^{-4})^2} = 105.12\Omega$$

E. The Circular Mil and American Wire Gage

In Europe, as among physicists everywhere, it has long been a practice to specify a particular size of copper wire by giving its diameter in millimeters. In Europe, wire is produced in sizes which are simple decimal fractions of 1mm, such as 0.1mm, 0.5mm, etc.

The engineering system still used in America was an attempt to simplify the calculations involved in determining the resistance of so many feet of wire by expressing the length of the wire in feet, its cross section area in *circular mils,* and its resistivity in ohm-circular mils per foot. In this system a *mil* is 0.001 inch. A *circular mil* is defined as an area equal to the area of a circle whose diameter is 1 mil. Thus, the cross-sectional area of a wire in circular mils is equal to the square of the wire diameter in mils:

$$A_{(circular\ mils)} = [d_{(mils)}]^2$$

(Equation 6-3)

The purpose of this system is to transfer most of the arithmetic (and particularly the difficult job of keeping track of the place of the decimal point), to a resistivity table, thereby making it much easier to perform resistance calculations with a slide rule.

With the popularization of the pocket calculator and the adoption of the metric system, this way of calculating the resistance of a length of wire no longer holds any mathematical advantage over the system using the mks units. Unfortunately, the circular mil is the basis for the American Wire Gage (AWG) wire size numbers. With this system, subtracting three from the AWG number of any particular wire size will give the AWG number of the wire size with twice the cross-sectional area and half the resistance for a given length of wire.

AWG Solid Wire (Copper or Tin-plated Copper)				
AWG Wire Size	Diameter in mm	Diameter in mils	Area in circular mils	Ohms per 1000 feet
30	.25	10	100	112.1
28	.32	12.6	159	72.9
26	.41	15.9	253	46.0
24	.50	20.1	404	28.3
22	.64	25.4	640	17.6
20	.81	32.0	1020	11.1
18	1.02	40.3	1620	6.92
16	1.29	50.8	2580	4.35
14	1.63	64.1	4110	2.72

Table 6-6

Table 6-6 gives the diameter in both mils and millimeters, the area in circular mils, and the resistance per thousand feet of some widely used AWG number copper wires.

It should be noted that the same AWG numbers are used for both solid and multi-strand wires of a given cross section, even though the stranded wire tends to show a slightly lower resistance.

1. *Temperature and resistance*

As pointed out in the previous paragraph, as well as in Section 5, the resistance of a conductor varies with the conductor temperature. In most common materials used in avionics equipment, and within the operating range of this equipment, the change in resistance per degree of temperature change remains fairly constant for each material.

This rate of change in resistance per centigrade degree change in temperature is called the *temperature coefficient of resistance* of the material and is usually symbolized by the Greek letter alpha (α). For the annealed copper normally used for electrical wiring, $\alpha = 0.00393$ around 20 °C.

If the resistance of a wire and the α of the material are known at any given temperature, the resistance at any other temperature may be calculated by using the following formula, even if the actual length of the wire is not known:

$$R_2 = R_1\ [1 + \alpha (t_2 - t_1)]$$

(Equation 6-4)

where R_2 is the resistance to be calculated, R_1 is the known resistance at temperature t_1, α is the temperature coefficient that applies to the range of temperatures in question ($\alpha = 0.00393$ is generally satisfactory for copper under normal operating conditions), and t_2 is the temperature at which we are trying to determine the new resistance.

It should be noted that the resistance of a copper wire *increases* as its temperature increases, as predicted by the discussion in sections 4 and 5. The temperature coefficients of resistance for some of the commonly used conductors is given in Table 6-7.

Material	Alpha α
Aluminum	+0.0039
Carbon	−0.0005
Constantan	+0.000002
Copper	+0.00393
Iron	+0.005
Nichrome	+0.0004
Silver	+0.0038

Table 6-7

In most applications, the change in wire resistance due to temperature change is unimportant. In certain measuring devices, however, alloys such as constantan, made up of 60% copper and 40% nickel, are used, since they possess a much smaller temperature coefficient than copper or silver.

F. Wire and Cable

In spite of the fact that wires and cables play a central role in all electronic equipment, surprisingly little is ever said about them in most training courses. What sort of wire or cable to use in a given situation is usually learned through experience, or from questioning "the old-timers," or "the hard way" by having to redo a job that has failed to pass a safety inspection. Although experience is indeed the best teacher, the following paragraphs are intended to provide some general information and hints with regard to the selection and use of wire and cable.

While the word *cable* is commonly used to mean any heavy conductor, in electronic usage, a *wire* is a *single* conductive path whether made of one solid piece of wire or of many fine metal threads twisted together, but a *cable* consists of *two or more* separate conductive paths bound into a single package.

1. Solid wire

Because vibration would quickly cause solid wires to "fatigue" and break, their use in airborne and even in ground equipment is discouraged, unless the total length is small and the wire is rigidly supported. Component leads, transformer coils, and rigid bus wires account for most of the solid wire used in avionics. The solid wire may be copper with a thin coating of tin to prevent corrosion, or solid copper with an enamel lacquer or plastic insulation bonded to the copper.

The choice of which size and what sort of wire to use for a given job depends on many factors. These include: how much current the wire is required to carry, the potential difference between the wire and nearby conductors, the operating temperature of the wire and of the area in which the wire will be, and the presence of hydraulic fluid, oil, water or other chemicals.

The leads which form parts of many components such as transistors, resistors, and capacitors are tin-, silver-, or gold-plated copper ranging in size from AWG #28 to AWG #12 for heavier current applications. Since these leads are always made as short as possible in practice, the effects of vibration are at a minimum. The presence of terminals or printed circuit copper strips also minimizes the heating effects, so that finer wires may carry fairly high currents.

In transformers and coils, however, where it is necessary to pack many loops or turns of wire in a small space, heat can become a problem, especially since it is necessary to electrically insulate the turns of wire from each other. The normal insulating material is a thin coat of enamel, which takes up less space than most other insulating materials.

To prevent overheating of the wire and charring or burning of the enamel insulation, it is necessary to use a wire that will not get hotter than about 80 — 100°C, while passing the expected maximum current. Normally this can be achieved by choosing wire with a cross-sectional area of around 800 to 1000 circular mils per ampere of current in large multi-layer coils and transformers.

Recently, the use of fluourethylene-based plastics in place of the familiar enamel insulation has made it possible to operate coils and transformers at higher temperature without the charring or burning of the wire insulation. This means that finer wire may be used, resulting in a size and weight reduction.

2. *Stranded wire*

Exclusive of component leads and rigid bus wires, most of the point-to-point connections in avionics equipment is made with plastic-covered, stranded, hook-up wire. The reason stranded wire, consisting of a number of fine tin-plated wires twisted together, is preferred over solid wire is easy to demonstrate. Obtain half-inch lengths of stranded and solid wire of the same diameter, AWG #18 or #22 for example. Starting with the solid wire, grasp both ends with pliers, begin bending and straightening the wire. Be sure to count the number of times the wire is bent. Eventually the wire will become easier to bend. This marks the onset of metal "fatigue." If one continues to bend and straighten the wire after the onset of fatigue, the wire will break. Record the number of times the wire was bent before it broke. If one repeats the experiment with the stranded wire, it will be seen how much longer it takes to break all the strands. Clearly, the stranded wire will stand up better against vibration and hard usage than the solid wire.

In addition to its increased strength, a further advantage of stranded wire over solid wire is its decreased resistance. Table 6-8 gives the AWG size, number of strands, strand diameter, cross-sectional area, and resistance per thousand feet of some stranded hook-up wire sizes. Except where the rigidity of solid wire is desired, as for self-supporting bus wires or component leads, and where the leads must support the component, military and federal specifications generally require the use of stranded wire for all hook-up and point-to-point wiring. The major disadvantage of stranded wire is that it is somewhat harder to work with than solid wire, and requires a bit more care in the preparation of wire ends for joining.

AWG wire size	Number of strands	Single strand diameter (mils)	Total area in circular mils	Ohms/ 1000 feet
30	3	6.3	119	93.6
28	7	5.0	175	70.0
26	7	6.3	278	49.0
24	7	8.0	448	28.4
22	7	10.0	700	19.0
20	7	12.6	1111	12.1
18	16	10.0	1600	7.89
16	19	11.3	2426	4.97
14	19	14.2	3831	3.13

Table 6-8

a. *Insulation*

It is a popular legend that when he was busy inventing the electric light bulb, Thomas Edison tried everything possible in order to obtain the proper filament material. Wire manufacturers, too, seem to have exhausted all the possibilities in their search for an effective insulating material to surround solid or stranded wires. Fortunately, the dozens of sorts of insulating material, each with its own special properties, advantages, and limitations, have largely been replaced by just a few materials.

The cheapest and most common of these is the familiar polyvinyl chloride (PVC) which, though it tends to soften and melt when heated to the temperatures required for soldering, is fire and chemical resistant, and generally sufficient for applications involving temperatures up to 100 °C and potential differences up to about 600 volts.

For more exacting uses, including temperatures of up to 200 °C, one of a number of fluorine-based plastic insulations may be used. Although more expensive, these offer superior heat resistance and less deformation or shrinkage when the wires are soldered.

Power, lighting, and control signal wires, particularly around airports, are subject to strict federal and military specifications. These wires are generally insulated with several layers of plastic and/or rubber insulation for extreme ruggedness and protection against the hazards of lightning strikes.

3. *Multi-conductor cable*

A multi-conductor cable is a group of separate, insulated wires formed into a bundle by an outer insulating jacket, and is used for connection in place of individual wires.

In some cables, individual conductors are paired by color-coding, so that each pair may constitute both the positive and negative leads of a complete circuit. In the electronics industry, cables containing up to sixty or more individual wires are commonly used, and much larger cables are found in telephone company installations.

Because it is impossible to replace an individual wire in a multi-conductor cable, equipment manufacturers prefer to use a wire harness. A wire harness is a bundle of individual wires tied together with plastic tie strips or laced together with nylon lacing thread.

A second type of multi-conductor cable that the avionics technician may expect to find more and more frequently as computer technology is applied to avionics, is the flat ribbon cable shown in Fig. 6-13. This type of cable has the advantage of providing the large number of individual wires required for the data, memory address, and control lines of the computer, while still retaining a degree of flexibility impossible in round multi-conductor cable.

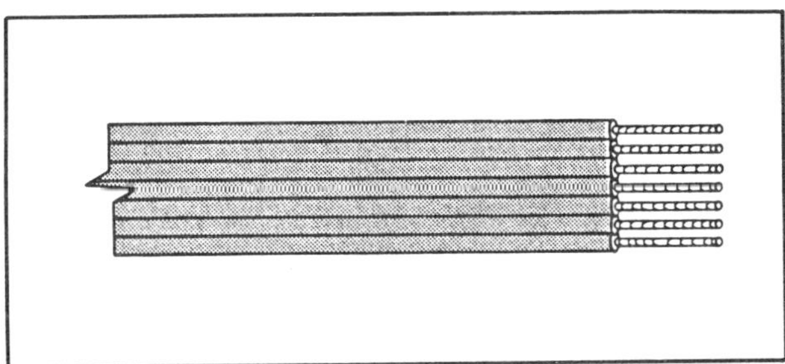

Fig. 6-13 As computer techniques become more common in avionics equipment, ribbon cable is more frequently used.

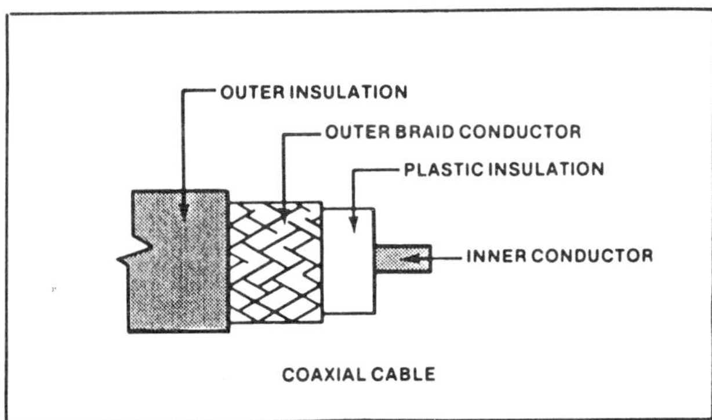

Fig. 6-14 Familiarly called "coax" in the aviation industry, coaxial cable is widely used for high frequency signals.

4. *Coaxial cable*

Coaxial cable, generally called "coax", is used wherever sensitive signal or control wires must be more than a few inches in length. As shown in Fig. 6-14, a coaxial cable consists of an inner wire, plastic insulation, a conductive braid shield, and an outer insulating shield. Since it is used primarily for the transmission of alternating current signals, the characteristics of coax will be discussed in the work on alternating current in this series.

A. *Self-test Questions:*

Fill in the blanks:

1. Most "ground side" equipment is designed to permit mounting in standard ________ ________ which require a panel width of ________ inches.
2. The ________ racking system is used for most airborne equipment.
3. The aluminum or plated steel body upon which electronic equipment is built is called a ________________.
4. The method of construction in which the components are connected by means of their leads or by separately connected hook-up wires is called ________ ______ ________ wiring.
5. A glass fiber or plastic board with bonded copper strips for electrical connection is called a ________ ________ ________.
6. A ________ hook-up wire might mark a ground connection in a vacuum tube or transistor circuit, while a red wire would probably be used for ________ or ________.
7. Fill in the resistance values symbolized by the following color bands:
   a. brown, black, red, gold ________
   b. green, blue, gold, silver, orange ________
   c. red, violet, orange, gold, yellow ________
   d. brown, black, blue ________
   e. white, brown, brown, red ________
8. A wire with a diameter of 16.3 mils has a cross-sectional area of ________ circular mils.

B. Problems for Thought and Calculation:

1. Describe coaxial cable. What might be the use of the outer braided shield?

2. Describe the American Wire Gage numbering system (include a definition of a circular mil).

3. A copper wire has a resistance of 54 ohms at room temperature (20 °C); what will its resistance be at 75 °C?

4. A carbon resistor has a resistance of 10 K ohms at 20 °C; what will its resistance be at 80 °C?

5. Calculate the resistance of 350 feet of AWG #26 wire at 75 °C. (Hint: calculate the resistance at 20 °C first.)

6. Calculate the resistance of 35 feet of copper wire with a diameter of 3 mils at 20 °C.

7. Discuss the advantages of a single, widely-accepted racking system such as the ATR system.

C. Experiments and Demonstrations:

1. Obtain a circuit board from an avionics unit. Try to read the color-coding of all the resistors on it.

2. Plan a visit to a commercial airline repair facility; notice the racking and connector systems used.How many different size units are visible? Are panel instruments sizes standardized?

SECTION VII

Fixed Resistors and Fixed Resistor Circuits

This section continues the introduction to electronic components and their applications which began in Section 6. Remember that in the last section, two problems were pointed out — that the standard color coding system for resistors has no way of coding more than two significant figures (there is no way, for example, to show a resistance of 15.6 KΩ using the color code); and that the rated power dissipation of resistors seems to be ignored.

In this section, it will be seen that neither of these problems is real. We will learn about the various kinds of *non-adjustable* or *fixed* resistors commonly used in avionics equipment and the properties of each kind. *Variable resistors* (resistors which can be adjusted to give the value of resistance required at a particular moment), and *thermistors* (resistors whose resistance is made to vary with temperature in a precise way), will be discussed in Section 8. A number of practical examples are also offered so the student will be familiar with the analysis and design of resistor networks.

A. *Standard Resistor Values*

In the early days of electronics, color coding and industry-wide standard values were unheard of; each manufacturer was able to sell his "little wonder" resistors as they came out of his factory, without worrying about tolerances or final value checking. For example, the early *pencil mark* grid leak resistors, used in vacuum tube radios, consisted of a carbon line drawn on a glass plate. The range of actual resistance must have been enormous, but if the component worked, that was enough.

As the electronics industry came of age, standardization of resistor values came into being along with the color coding system. The object of standardizing certain resistance values was to reduce the number of resistors which manufacturers had to produce, as well as the number that had to be stocked for replacement. This was accomplished by carefully choosing values so that it would be possible to purchase a 10% tolerance resistor whose nominal or marked resistance would be within 10% of nearly any resistance that might be required, or, a 5% tolerance resistor whose nominal or marked resistance value would be within 5% of any resistance that might be required. This is accomplished by choosing a number of base values. The resistors actually produced have resistance values which are decimal multiples (×1, ×10, ×100, etc.) of the base values. The base values of 10% resistors that provide the most efficient system are those listed along the side of Table 7-1. The 10% tolerance resistors manufactured are decimal multiples of these base values, as shown along the top row of the table. The *nominal* resistance value of any standard 10% tolerance resistor, (that is, the center range resistance value which is shown by the color code), may be found by tracing across the proper base number row and down the proper decimal multiplier column. The resistance range values, listed next to each nominal resistance value in the table, give the range over which a new 10% tolerance resistor of that nominal value may vary. For example, a 12 KΩ, 10% resistor may have an actual resistance as high as 13.2 KΩ or as low as 10.8 KΩ.

The range values in the table are also useful for determining which standard value resistor to use after the required value for a circuit has been calculated. For example, in a particular circuit we have a current flow of 40 mA and wish to insert a resistor that will provide a potential difference of 7 volts when this current flows through it. The calculation of the required resistance is simple, using Ohm's law $R = E/I = 175\Omega$.

Suppose we wish to use a 10% tolerance resistor. From Table 7-1 we see that 175Ω lies within the range of a standard 180Ω, 10% unit. As for the power dissipation of the resistor, using $P = I^2R$, we get a required dissipation of 0.28 watts, slight-

10% RESISTORS STANDARD VALUES

Base Number	Decimal Multiplier: 10^{-2} silver — Resistor Value Ohms	10^{-2} silver — 10% Tolerance Range	10^{-1} gold — Resistor Value Ohms	10^{-1} gold — 10% Tolerance Range	10^{0} black — Resistor Value Ohms	10^{0} black — 10% Tolerance Range	10^{1} brown — Resistor Value Ohms	10^{1} brown — 10% Tolerance Range	10^{2} red — Resistor Value Ohms	10^{2} red — 10% Tolerance Range	10^{3} orange — Resistor Value Ohms	10^{3} orange — 10% Tolerance Range	10^{4} yellow — Resistor Value Ohms	10^{4} yellow — 10% Tolerance Range	10^{5} green — Resistor Value Ohms	10^{5} green — 10% Tolerance Range
10 brown black	0.10	0.09-0.11	1.0	0.9-1.1	10	9-11	100	90-110	1K	0.9K-1.1K	10K	9K-11K	100K	90K-110K	1M	0.9-11.M
12 brown red	0.12	0.108-0.132	1.2	1.08-1.32	12	10.8-13.2	120	108-132	1.2K	1.08K-1.32K	12K	10.8K-13.2K	120K	108K-132K	1.2M	1.08M-1.32M
15 brown green	0.15	0.135-0.165	1.5	1.35-1.65	15	13.5-16.5	150	135-165	1.5K	1.35K-1.65K	15K	13.5K-16.5K	150K	135K-165K	1.5M	1.35M-1.65M
18 brown gray	0.18	0.162-0.198	1.8	1.62-1.98	18	16.2-19.8	180	162-198	1.8K	1.62K-1.98K	18K	16.2K-19.8K	180K	162K-198K	1.8M	1.62M-1.98M
22 red red	0.22	0.198-0.242	2.2	1.98-2.42	22	19.8-24.2	220	198-242	2.2K	1.98K-2.42K	22K	19.8K-24.2K	220K	198K-242K	2.2M	1.98M-2.42M
27 red violet	0.27	0.243-0.297	2.7	2.43-2.97	27	24.3-29.7	270	243-297	2.7K	2.43K-2.97K	27K	24.3K-29.7K	270K	243K-297K	2.7M	2.43M-2.97M
33 orange orange	0.33	0.297-0.363	3.3	2.97-3.63	33	29.7-36.3	330	297-363	3.3K	2.97K-3.63K	33K	29.7K-36.3K	330K	297K-363K	3.3M	2.97M-3.63M
39 orange white	0.39	0.351-0.429	3.9	3.51-4.29	39	35.1-42.9	390	351-429	3.9K	3.51K-4.29K	39K	35.1K-42.9K	390K	351K-429K	3.9M	3.51M-4.29M
47 yellow violet	0.47	0.423-0.517	4.7	4.23-5.17	47	42.3-51.7	470	423-517	4.7K	4.23K-51.7K	47K	42.3K-51.7K	470K	423K-517K	4.7M	4.23M-5.17M
56 green blue	0.56	0.504-0.616	5.6	5.04-6.16	56	50.4-61.6	560	504-616	5.6K	5.04K-6.16K	56K	50.4K-61.6K	560K	504K-616K	5.6M	5.04M-6.16M
68 blue gray	0.68	0.612-0.748	6.8	6.12-7.48	68	61.2-74.8	680	612-748	6.8K	6.12K-7.48K	68K	61.2K-74.8K	680K	612K-748K	6.8M	6.12M-7.48M
82 gray red	0.82	0.738-0.902	8.2	7.38-9.02	82	73.8-90.2	820	738-902	8.2K	7.38K-9.92K	82K	73.8K-90.2K	820K	738K-902K	8.2M	7.38M-9.02M

TABLE 7-1

5% RESISTORS STANDARD VALUES
Decimal Multiplier

Base Number	10^{-2} Resistor Value Ohms	5% Tolerance Range ohms	10^{-1} Resistor Value Ohms	5% Tolerance Range ohms	10^{0} Resistor Value Ohms	5% Tolerance Range ohms	10^{1} Resistor Value Ohms	5% Tolerance Range ohms
10	.1	.095-.105	1	.95-1.05	10	9.5-10.5	100	95-105
11	.11	.105-.115	1.1	1.05-1.15	11	10.5-11.5	110	105-115
12	.12	.114-.126	1.2	1.14-1.26	12	11.4-12.6	120	114-126
13	.13	.124-.136	1.3	1.24-1.36	13	12.4-13.6	130	123.5-136
15	.15	.142-.157	1.5	1.42-1.57	15	14.2-15.7	150	142-157
16	.16	.152-.168	1.6	1.52-1.68	16	15.2-16.8	160	152-168
18	.18	.171-.189	1.8	1.71-1.89	18	17.1-18.9	180	171-189
20	.20	.19-.21	2.0	1.9-2.1	20	19-21	200	190-210
22	.22	.209-.231	2.2	2.09-2.31	22	20.9-23.1	220	209-231
24	.24	.228-.252	2.4	2.28-2.52	24	22.8-25.2	240	228-252
27	.27	.256-.283	2.7	2.56-2.83	27	25.6-28.3	270	256-283
30	.30	.285-.315	3.0	2.85-3.15	30	28.5-31.5	300	285-315
33	.33	.313-.346	3.3	3.13-3.46	33	31.3-34.6	330	313-346
36	.36	.342-.378	3.6	3.42-3.78	36	34.2-37.8	360	342-378
39	.39	.370-.409	3.9	3.70-4.09	39	37.0-40.9	390	370-409
43	.43	.408-.451	4.3	4.08-4.51	43	40.8-45.1	430	408-451
47	.47	.446-.493	4.7	4.46-4.93	47	44.6-49.3	470	446-493
51	.51	.484-.535	5.1	4.84-5.35	51	48.4-53.5	510	484-535
56	.56	.532-.588	5.6	5.32-5.88	56	53.2-58.8	560	532-588
62	.62	.589-.651	6.2	5.89-6.51	62	58.9-65.1	620	589-651
68	.68	.646-.714	6.8	4.64-7.14	68	46.4-71.4	680	646-714
75	.75	.712-.787	7.5	7.12-7.87	75	71.2-78.7	750	712-787
82	.82	.779-.861	8.2	7.79-8.61	82	77.9-86.1	820	779-861
91	.91	.864-.955	9.1	8.64-9.55	91	86.4-95.5	910	864-955

Base Number	10^{2} Resistor Value Ohms	5% Tolerance Range ohms	10^{3} Resistor Value Ohms	5% Tolerance Range ohms	10^{4} Resistor Value Ohms	5% Tolerance Range Ohms	10^{5} Resistor Value Ohms	5% Tolerance Range Ohms
10	1K	950K-1.05K	10K	9.5K-10.5K	100K	95K-105K	1M	.95M-1.05M
11	1.1K	1.05K-1.15K	11K	10.5K-11.5K	110K	105K-115K	1.1M	1.05M-1.15M
12	1.2K	1.14K-1.26K	12K	11.4K-12.6K	120K	114K-126K	1.2M	1.14M-1.26M
13	1.3K	1.24K-1.36K	13K	12.3K-13.6K	130K	124K-136K	1.3M	1.24M-1.36M
15	1.5K	1.42K-1.57K	15K	14.2K-15.7K	150K	142K-157K	1.5M	1.42M-1.57M
16	1.6K	1.52K-1.68K	16K	15.2K-16.8K	160K	152K-168K	1.6M	1.52M-1.68M
18	1.8K	1.71K-1.89K	18K	17.1K-18.9K	180K	171K-189K	1.8M	1.71M-1.89M
20	2K	1.9K-2.1K	20K	19K-21K	200K	190K-210K	2M	1.9M-2.1M
22	2.2K	2.09K-2.31K	22K	20.9K-23.1K	220K	209K-231K	2.2M	2.09M-2.31M
24	2.4K	2.28K-2.52K	24K	22.8K-25.2K	240K	228K-252K	2.4M	2.28M-2.52M
27	2.7K	2.56K-2.83K	27K	25.6K-28.3K	270K	256K-283K	2.7M	2.56M-2.83M
30	3K	2.85K-3.15K	30K	28.5K-31.5K	300K	285K-315K	3M	2.85M-3.15M
33	3.3K	3.13K-3.46K	33K	31.3K-34.6K	330K	313K-346K	3.3M	3.13M-3.46M
36	3.6K	3.42K-3.78K	36K	34.2K-37.8K	360K	342K-378K	3.6M	3.42M-3.78M
39	3.9K	3.7K-4.09K	39K	37K-40.9K	390K	370K-409K	3.9M	3.7M-4.09M
43	4.3K	4.08K-4.51K	43K	40.8K-45.1K	430K	408K-451K	4.3M	4.08M-4.51M
47	4.7K	4.46K-4.93K	47K	44.6K-49.3K	470K	446K-493K	4.7M	4.46M-4.93M
51	5.1K	4.84K-5.35K	51K	48.4K-53.5K	510K	484K-535K	5.1M	4.84M-5.35M
56	5.6K	5.32K-5.88K	56K	53.2K-58.8K	560K	532K-588K	5.6M	5.32M-5.88M
62	6.2K	5.89K-6.51K	62K	58.9K-65.1K	620K	589K-651K	6.2M	5.89M-6.51M
68	6.8K	6.46K-7.14K	68K	64.6K-71.4K	680K	646K-714K	6.8M	6.46M-7.14M
75	7.5K	7.12K-7.87K	75K	71.2K-78.7K	750K	712K-787K	7.5M	7.12M-7.87M
82	8.2K	7.79K-8.61K	82K	77.9K-86.1K	820K	779K-861K	8.2M	7.79M-8.61M
91	9.1K	8.64K-9.55K	91K	86.4K-95.5K	910K	864K-955K	9.1M	8.64M-9.55M

TABLE 7-2

ly above the capabilities of a 1/4-watt unit, so a 1/2-watt resistor would be used.

Since there are twelve base numbers in Table 7-1, each power of 10 from 10^{-2} to 10^{5} results in 12 resistors, or about 96 different 10% tolerance resistor values which are regularly found in use. It should also be noted that two significant figures and a decimal multiplier are enough to specify any of the standard resistor values; this is why the standard resistor color code does not need to deal with values like 13.4 KΩ. (13.4 KΩ, by the way, is one of the resistance values that does not fall within ±10% of a standard value 10% tolerance resistor.)

Because of the reduced range of 5% tolerance resistors, twenty-four values are required for each power of ten decimal multiplier. This means that a total of 192 different 5% tolerance standard resistor values must be stocked to cover the most frequently used resistors, as shown in Table 7-2.

Most of the resistors used in avionics equipment are required to be 5% tolerance. Thus, although they are slightly more expensive than 10% tolerance components, they are stocked by most avionics repair facilities instead of the 10% resistors. Again, it should be noted that a color code using two significant figures plus a decimal multiplier will be sufficient to encode the standard 5% tolerance values.

When resistors of greater precision than 5% tolerance are required, 2%, 1%, and even 0.5% tolerance resistors are used. Although resistors of such high precision are not as common as the standard value 5% tolerance units, they are by no means rare in avionics equipment. In these cases, resistance values are printed on the resistors, rather than being color-coded. This is because values with three, and even four significant figures are possible in 0.5% tolerance resistors. In certain cases, color-coded, standard value resistors with tolerance ratings of 2% or 1% are used when the equipment design demands it and when only a small deviation from the nominal resistance can be tolerated.

A typical application of 2% tolerance standard value resistors is found in the monitoring units of an instrument landing system. These monitoring units must separate the various parts of the complex landing system signals being transmitted to the airplane approaching an airport and constantly check that each "signal part" is being transmitted with the correct power and in the proper relationship to the other parts. Separation of these signal parts is an extremely difficult affair, since several of them are quite similar. This task is made possible through the use of highly sophisticated *selecting circuits* using precision resistors and capacitors of 2% or 1% tolerance.

B. Resistor Power Dissipation and Physical Size

It may come as something of a surprise to learn that the size of a resistor is totally independent of its resistance. Today, resistor manufacturers are able to modify their materials in order to place a resistance of 0.1Ω in the same size component as a resistance of 100 MΩ. The physical size of a resistor depends only on the amount of heat that the resistor will be required to dissipate while operating, which is a function of resistor surface area.

The larger the resistor, the more heat it is able to dissipate. As discussed in Section 5, the power dissipated in a resistor is converted into heat. Regardless of its resistance value, a large resistor, which will dissipate more heat, will have a greater power rating than a physically smaller resistor, which will burn up because its smaller surface area is not able to dissipate as much heat as the larger resistor. This is the reason it is not necessary to use the color code or any other special marking to encode the power dissipation of resistors rated at two watts or less. The size of the resistor alone (which is pretty well standardized throughout the industry), is sufficient to denote which of the standard power ratings the resistor has.

Fig. 7-1 gives the dimensions of resistors with standard power ratings based on an operating temperature of 70°C (158°F). For resistors with a power dissipation, or wattage rating of 3 watts or greater, size and shape are not standardized, and the wattage rating, like the resistance, is printed on the component.

Due to the trend toward smaller equipment and to the reduced power requirements of modern circuits, 1- and 2-watt resistors are becoming less common in new equipment. Even the 1/2-watt resistor, which was the most common component in

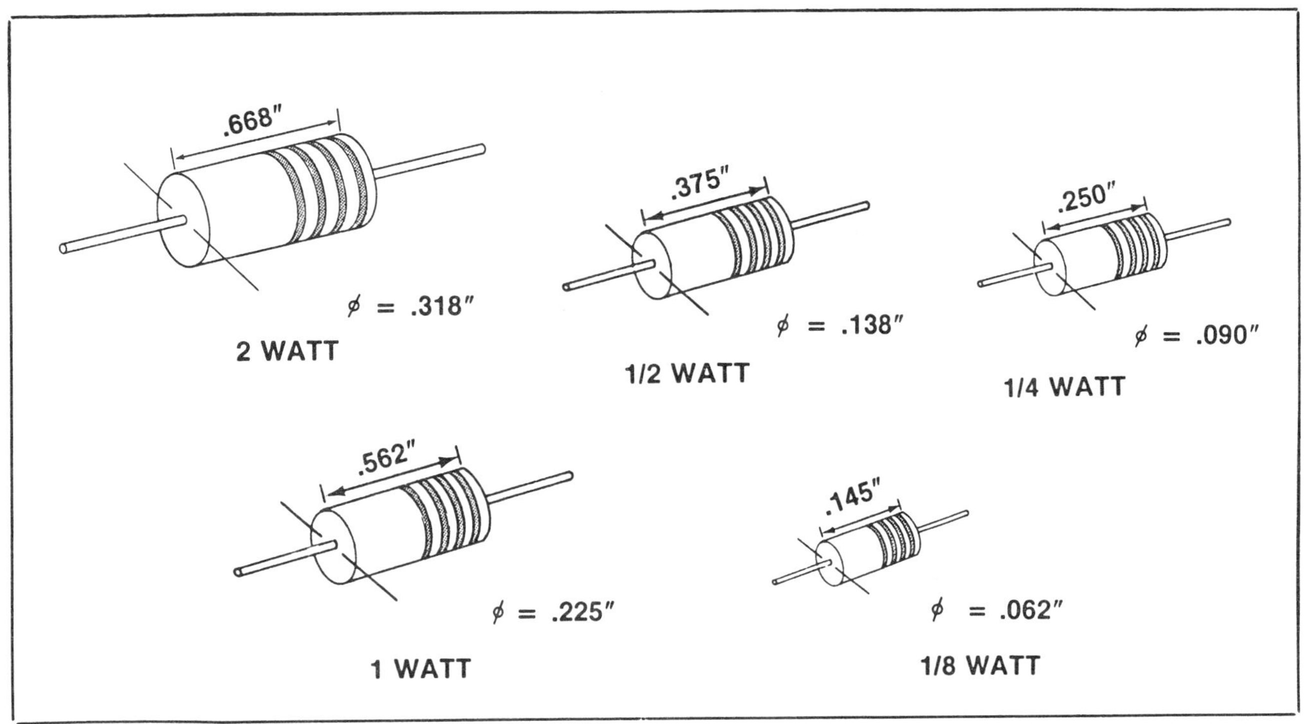

Fig. 7-1 Carbon composition resistor sizes and power dissipation ratings.

vacuum tube circuits, is now less common than the 1/4-watt size. In the near future, with the increasing application of computer techniques to avionics equipment design, the tiny 1/8- and 1/10-watt resistors may be expected to become the most widely used.

C. *Fixed Resistor Types*

1. *Wire-wound resistors*

Since the earliest days of the components industry, there have been two broad categories of fixed resistors, resistors whose resistance value is not adjustable. One type consists of a length of wire with fairly high resistivity, such as *nichrome*, wound on a core of heat-resistant material, often ceramic. This early form of resistor is still the last word, in fact, it is the only word for resistors with a power dissipation rating of over 3 watts to a maximum of 250 watts. Fig. 7-2 shows the construction details of typical wire-wound units.

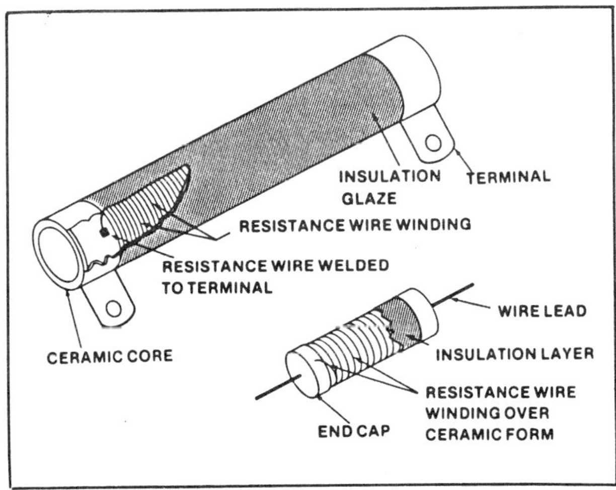

Fig. 7-2 Two typical wire-wound resistors.

Because wire with precisely formulated characteristics is used in the construction of wire-wound resistors, they can be designed for a wide range of uses. Some requirements met by wire-wound resistors are extremely low resistance, small resistance change over a wide range of operating temperature, and extreme precision to a tolerance of 0.05%. However, high cost and problems in handling high frequency AC are two major drawbacks to the use of wire-wound resistors.

2. *Carbon composition resistors*

Carbon resistors were also used in the early days of the electronics industry. One form of carbon resistor consisted of a carbon rod of a specific length and thickness with connecting wires wrapped about either end. This primitive device gave rise to the today's *carbon composition resistor.*

There are several ways of manufacturing carbon composition resistors. One manufacturing process, the most popular so far, consists of embedding lead wires in a small cylindrical carbon composition slug which has been formulated to give the required resistance. (Standard resistances from 1Ω to 22MΩ are usually stocked by suppliers.) The whole unit is then sealed with a plastic jacket and color-code bands are painted on. (Fig. 7-3) Carbon composition resistors are so cheap, costing the volume purchaser only pennies per unit, that literally billions of them are manufactured and used every year, in spite of the numerous drawbacks they present.

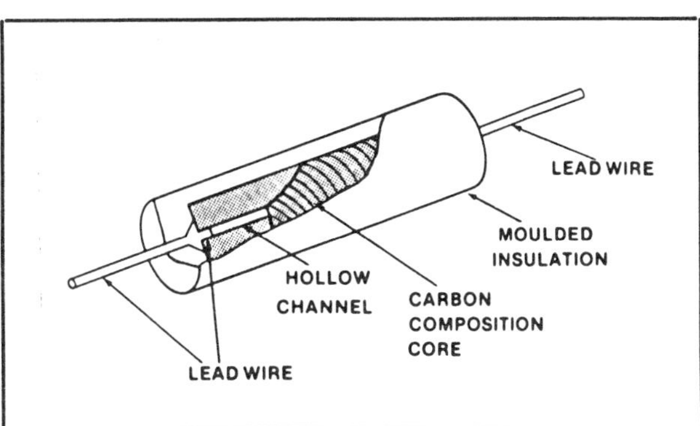

Fig. 7-3 Carbon composition resistor with one type of internal structure shown.

There are a number of disadvantages associated with carbon composition resistors. First, they are surprisingly sensitive and can be easily damaged. For example, carbon composition resistor power ratings should never be exceeded. When a 1/2-watt carbon composition resistor is used to dissipate a full watt, even for the brief instant at initial equipment turn-on, until steady state conditions are reached, the heating effect will considerably shorten the life of the component, and will also cause a change in resistance value. Second, moisture and temperature shock during construction and operation can also shorten the effective life of carbon composition resistors and cause their resistance value to change.

Further, even if operated carefully, a carbon composition resistor will show a steady change in resistance value throughout its life span. This gradual change can result in a variation in resistance value of 15% or so, even for a 5% tolerance unit. In other words, a 100Ω 5% resistor, which originally had a resistance range of 100Ω±5%, that is, 95 to 105Ω, might end its life with a resistance range of 80.75Ω to 120.75Ω, or a tolerance of 20%. In certain circuits, changes of this sort could cause a gradual deterioration of performance, faulty operation, or even failure of other components. In critical circuits, where a *worst-case* shift of this sort could cause trouble (note that in most instances the value change of the carbon composition resistor will not be as drastic), either a wire-wound or one of the film type resistors discussed below must be used. In many circuits, carbon composition resistors are sufficient, for they are quite reliable if they are not abused.

The third major difficulty with carbon composition resistors is that they are not suitable for high voltage or high-frequency, low-noise circuits. This becomes an important consideration in tuned circuits, such as those used in transmitters and receivers. One of the characteristics of a tuned circuit is that the voltage across it may be quite high, even though the equipment power supply may be a low-voltage source. Tuned circuits are generally required to be as noiseless as possible, that is, they are not supposed to hash up the impressed signal or produce the hissing sound that sometimes is heard when listening to a weak radio station. Because of this, tuned circuits use special high-voltage or low-noise film resistors.

3. *Film resistors*

The film resistor differs from a carbon composition resistor in that the resistance element is usually a thin film or coating of metal deposited on an insulating core of ceramic or other material. As shown in Fig. 7-4, this coating is then etched

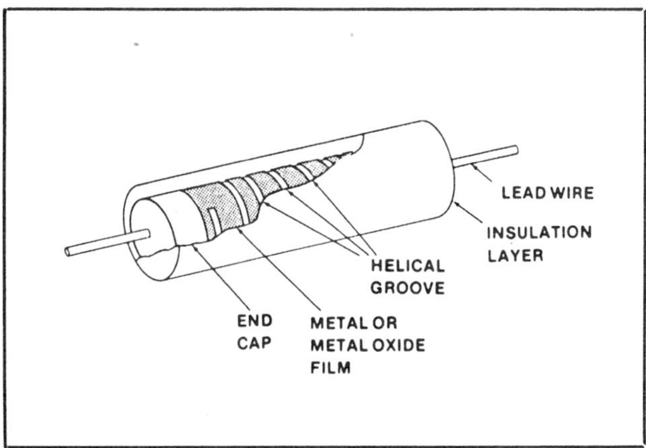

Fig. 7-4 A film resistor with typical internal structure shown.

or cut in a *helix*, thereby increasing the length of the path between the resistor leads. The leads are attached to the core by means of metal caps and the whole unit is covered, like the carbon composition resistor, with an epoxy or other insulating plastic jacket.

Film resistors are regularly produced in a greater range of power dissipation ratings than carbon composition resistors. The standard ratings are: 1/20-, 1/10-, 1/8-, 1/4-, 1/2-, 1-, and 2-watt units. They are more stable than carbon resistors, with a smaller percent change in resistance value resulting from temperature change or operational life. Also, since the coating process can be very precisely controlled, film resistor tolerances are better than those obtainable in carbon composition resistors. Standard film resistors are available in 5%, 2%, 1%, 0.5% and 0.1% tolerances.

The principle drawbacks of film resistors is the difficulty of obtaining extremely low and extremely high range of resistances (the standard range is 10 ohms to 10 megohms). Also the component cost is relatively high, being twenty to forty times that of carbon composition resistors.

D. Fixed Resistor Circuits

1. General hints on the use and replacement of fixed resistors

As mentioned previously, fixed resistors are reliable devices if properly handled, and will function for a long time in properly designed circuits. Unlike motors, batteries, or vacuum tubes, there is no functioning part in a resistor that will wear out or become used up.

Sometimes, however, a resistor will fail completely, generally because the failure of another component or short-circuit will cause an excessive amount of current to flow through the resistor, causing it to burn up. In general, burned resistors are relatively easy to spot. Look for charred or cracked surfaces, and blistered or discolored color-code bands. Once the outer insulation of a resistor has become cracked or scorched, the component must be replaced. Even though the part may be working, its life expectancy has been seriously shortened, and its resistance value has probably been severely changed.

Federal regulations require careful attention in the selection of replacement parts. A burned-out film resistor must be replaced by a film resistor of equal resistance and tolerance, equal power dissipation, and equal or better reliability. Usually, carbon composition types cannot be used in place of burned out film or wire-wound types, although film resistors *may* be permitted in place of carbon compositions, if the characteristic resistance, tolerance, and power dissipation ratings are equal. In the case of wire-wound resistors, it is especially wise to use a new unit of exactly the same type and manufacturer.

When servicing equipment which has malfunctioned, the operating temperatures of the various resistors in the unit may provide clues to the reason for the malfunction. Although resistors can be quite hot in some circuits in normal operation, excessive temperature, evidenced by the special and easy-to-recognize smell of a burning carbon composition resistor, is a sure sign of some failure in that particular circuit. Keep in mind that one- and two-watt resistors are used in circuits to dissipate a fairly large amount of power. These components can be expected to run warm to hot during operation. The same is true for wire-wound types. On the other hand, precision resistors with a tolerance of 2% or better, and 1/8-watt resistors of all types, should be expected to run cool, since they are *generally* not expected to dissipate much power.

In all cases where the technician may be called upon to build a new circuit, resistors must be selected so as to provide a power dissipation safety margin above the calculated *maximum* power dissipation of the resistor in normal operation in the circuit.

2. Voltage dropping resistors

In the first section of this book, it was pointed out that avionics circuits can be understood as ways of controlling and applying energy to do the work of communicating information to or controlling an aircraft. This electrical energy must be distributed to different sorts of loads, and the amount of energy applied to each load must be strictly controlled to avoid damage.

One of the most common methods of doing this consists of limiting the potential difference or voltage applied across a load. This is very often done by inserting a suitable resistor in series with the load.

For example, let us assume that we have an integrated circuit which is designed to draw a current of 5 mA and to operate when a potential difference of 5.1 volts DC is connected across it. Further, assume that we have a 27-volt DC power supply available in the area in which this component is to be installed. In order to operate the integrated circuit from this power supply, we must obtain a potential difference of 5.1 volts DC and limit the current through the load to approximately 5 mA. The simplest solution to this problem is to use a voltage dropping resistor in *series* with the integrated circuit. (The use of resistors in parallel with the load to limit the current through the load, is an arrangement commonly called a *shunt*, will be discussed in a later paragraph.)

Calculation of the proper series resistor is fairly simple in this case. We know the current through the resistor, for it will be 5 mA, the same as the current required by the integrated circuit. The voltage drop across the resistor will have to be the voltage supplied by the power supply minus the potential difference required across the integrated circuit, or 27 − 5.1 = 21.9 volts DC. Using Ohm's law, the required resistance would be 4,380Ω. The amount of power that this resistance would dissipate in the circuit would be $P = I^2R = 0.11$ watt.

Once these theoretical values have been calculated (Fig. 7-5), the work of designing a practical circuit using real components may begin. The first step is to determine the tolerances or degree of precision needed. This depends upon the requirements of the integrated circuit as shown in the manufacturer's specifications. The recommended operating voltage range for this integrated circuit may show a range of 4.5 to 5.5 volts DC, and a typical current range from 4.9 to 5.1 mA. In the interest of reliability, we will want to provide an operating voltage and current as close to the middle of the recommended range as possible, and we will select our voltage dropping series resistor accordingly.

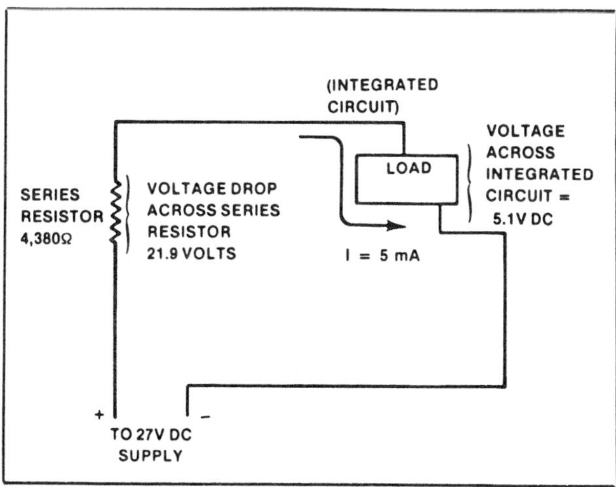

Fig. 7-5 A voltage-dropping resistor placed in series with a load.

Of course, we could install a 1% or 0.5% tolerance resistor with a nominal value of 4.38 KΩ, but the expense may not be necessary. Let us look at the 10% tolerance resistance range given in Table 7-1. The value 4.38 KΩ lies within the range of a standard 4.7 KΩ resistor, but will this resistor value meet the requirements of the integrated circuit?

A 4.7 KΩ, 10% resistor may have an actual resistance of between 4,230Ω and 5,170Ω. The integrated circuit, which required between 4.9 and 5.1 mA at a voltage of 5.0 volts may be thought of as a resistance with a range of 9870.4Ω (min.) to 1020.4Ω (maximum). In building this circuit we may by chance get any one of four maximum-minimum combinations for the two components involved (Table 7-3). This table shows that when the series resistor has a resistance value at the maximum of its range, the potential difference across the integrated circuit will not be within the recommended range. As a result of these calculations, we can see that we cannot safely use a 10% tolerance resistor in this circuit.

If we try a 5% tolerance resistor the situation changes. As we see from Table 7-2, a standard 4.3 KΩ, 5% resistor closely approaches the value calculated for the series resistor. The resistance range for a nominal value 4.3 KΩ, 5% resistor runs from a minimum of 4085Ω to a maximum of 4515Ω. A charting of the possible combinations for the two series components is provided in Table 7-4. In each of the possible cases in Table 7-4, the potential difference across the integrated circuit is within the operating values recommended by the manufacturer.

This represents a usable circuit; the slightly excessive circuit current that flows when the series

10% Series Resistor and Integrated Circuit

Series Resistor Resistance In Ohms	Integrated Circuit Resistance In Ohms	Total Current In Series Circuit In Amperes	Voltage Drop Across Series Resistor In Volts	Potential Difference Across Integrated Circuit In Volts
4230 (minimum)	980.4 (minimum)	0.0052	21.92	5.08 (O.K.)
4230 (minimum)	1020.4 (maximum)	0.0051	21.75	5.25 (O.K.)
5170 (maximum)	980.4 (minimum)	0.00438	22.70	4.30 (too low)
5170 (maximum)	1020.4 (maximum)	0.00436	22.55	4.45 (too low)

TABLE 7-3

resistance is at its minimum is assumed to have no effect on the integrated circuit. Should it be necessary to control the circuit current flow more precisely or to provide a safety margin against resistor value change, a 4.3 KΩ, 2% resistor might be used.

Let us assume that the nature of the components involved permits us to use a 4.3 KΩ, 5% carbon composition resistor as the series resistor. The maximum power dissipated by this resistor will be $P = I^2R = 0.11$ watt in the first example shown on Table 7-4. Since this is the worst case as far as power dissipation is concerned, a resistor capable of handling this comfortably will be adequate for all other cases. Allowing a safety factor, one would probably choose a 1/4-watt resistor (0.25 watt) rather than a 1/8-watt (0.125-watt) unit.

3. *Shunt resistors*

A second common practical circuit involving fixed resistors is the *current shunt circuit.* The shunt resistor is placed in parallel with the load. When the voltage dropping resistor reduces the potential difference across the load the shunt resistor acts to reduce the current through the load.

A common example of a shunt resistor is the meter shunt which is frequently used in parallel with an ammeter for in-circuit current measurements. The circuit of Fig. 7-6 represents a typical meter shunt circuit. Load L_1 is the power amplifier stage of a radio beacon transmitter, with an effective resistance of 50Ω connected to a 27 volt DC battery. The current given by Ohm's law is 540 mA.

5% Series Resistor and Integrated Circuit

Series Resistor Resistance In Ohms	Integrated Circuit Resistance In Ohms	Total Current In Series Circuit In Amperes	Voltage Drop Across Series Resistor In Volts	Potential Difference Across Integrated Circuit In Volts
4085 (minimum)	980.4 (minimum)	0.0053	21.77	5.23 (O.K.)
4085 (minimum)	1020.4 (maximum)	0.00528	21.60	5.40 (O.K.)
4515 (maximum)	980.4 (minimum)	0.0049	22.18	4.82 (O.K.)
4515 (maximum)	1020.4 (maximum)	0.00487	22.02	4.98 (O.K.)

TABLE 7-4

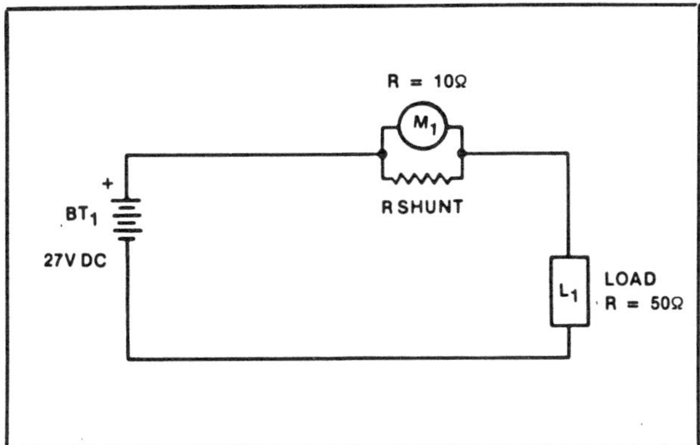

Fig. 7-6 A typical meter shunt. Most of the circuit currents passes through the shunt.

Assume that we wish to place an ammeter, M_1, in series with this circuit to monitor the current. The meter specified is a unit whose needle indicates full scale when a current of 10 mA is flowing through it, but a shunt resistor will be placed across the meter so that it registers half scale when a curent of 500 mA is flowing through the circuit. The resistance of the meter coil is 10Ω as shown. How would one go about selecting the proper resistor for the job? What is desired is that if the current through the circuit is 500 mA, ṁA will flow through the meter and 495 mA will flow through the shunt. Since the shunt resistor and the meter are in parallel, the voltage drop across them must be the same, or:

$V_{drop} = IR = 0.0005 \times 10 = 0.005$ volts in the case of the meter. This then must be equal to

$$0.4995 \times R_{shunt}$$

in the case of the shunt. Solving for R_{shunt} we get

$$R_{shunt} = 0.101\Omega.$$

If the calculated value is accepted, the required shunt would have to be a 1% precision resistor with a very small nominal value. Fortunately, this is not really required. Let us calculate the difference in the meter and shunt currents if a standard 0.1Ω, 1% resistor is used in place of the 0.101Ω calculated value.

If a 0.101Ω shunt is used, the combined resistance of the meter and the shunt will be 0.0999Ω. Putting this meter and shunt combination in series with L_1 will raise the total circuit resistance to 50.0999Ω and lower the circuit current to 538.9 mA. The current through the shunt is then given by:

$$\frac{R_{meter}}{R_{meter} + R_{shunt}} \times I_{total} = \frac{10}{10.101} \times 0.5389$$

$= 0.5345$ A.

The current through the meter is given by:

$$\frac{R_{shunt}}{R_{meter} + R_{shunt}} \times I_{total} = \frac{0.101}{10.101} \times 0.5389$$

$= 0.00538$ A or 5.38 mA.

If a standard value 0.1Ω 1% resistor is substituted for the 0.101Ω shunt, the effective range of this resistor will be between 0.101Ω and 0.099Ω. In the worst case condition, with a shunt resistor of 0.099Ω, the combined resistance of the shunt and meter in parallel will be 0.098Ω and the total circuit resistance will be 50.098Ω. Circuit current will be 0.5389 A or 538.9 mA, approximately the same as in the case of the meter and 0.101Ω shunt. The current through the shunt is:

$$\frac{R_{meter}}{R_{meter} + R_{shunt}} \times I_{total} = 0.00529 \text{ A or } 5.29 \text{ mA.}$$

The two meter currents, even in this worst case circumstance, differ by less than 2%, an accuracy sufficient for this kind of circuit. A 0.1Ω 1% resistor would be the logical choice. Since the power dissipated in the shunt is quite small, 0.028 watt, even a 1/10-watt unit would be sufficient, although a 1/4 resistor with its larger surface area would probably provide a more constant operating temperature regardless of the current. If the shunt should happen to fail, by opening, the entire current in the circuit

$$I = \frac{27 \text{ volts}}{10 + 50 \text{ ohms}} = 0.45\text{A}$$

would flow through the meter, destroying it. The shunt, therefore, must be extremely reliable and able to withstand mistreatment. A wire-wound unit would be a good choice.

E. The Voltage Divider

The voltage divider circuit previously introduced is a useful circuit when it is necessary to obtain two or more potential differences from a single source. As pointed out earlier, the circuits pictured in Figs. 4-11 and 4-13 are adequate as long as little or no current is drawn from the divider. However, since no allowance has been made for branch currents, the divider resistance values shown will not be satisfactory when branch currents are drawn.

To illustrate this, assume that currents of 1 A are required for loads connected to the output terminals of the divider of Fig. 4-13. As shown in Fig. 7-7, each of these loads may be thought of as a resistance connected in parallel with a portion of the voltage divider. In the absence of loads connected across the voltage divider, it is evident that all four resistors making up the divider would have to be of equal resistance, although the actual resistance itself would have no effect upon the voltage at each terminal.

As an exercise, assume that R_1 through R_4 are 100Ω resistors. Calculate the circuit current and voltage drop across each resistor. Notice that although the current through the resistors is different, the voltage drop across each resistor is equal to 3 volts in both cases. This is true because resistance and current in a parallel circuit are inversely proportional. This means that as we increase one, the other decreases. As long as each of the four resistances represents one-quarter of the total resistance in the circuit, the voltage drop across each one will remain one-quarter of the potential difference across the poles of the battery.

As soon as we connect a current-drawing load across one or more of the resistors in the divider, we change matters considerably. The same current *does not* pass through each of the resistors making up the divider. The current through R_1 in Fig. 7-7 is one ampere more than that through R_2. The voltage drops across the divider resistors are now no longer equal to each other even if the resistances remain equal.

In order to make the voltage drop across each resistor in this voltage divider equal to 3 volts, it is necessary to change the resistance of each resistor in such a way as to compensate for the differences in current through the resistors. The best way to do this is to begin with the divider

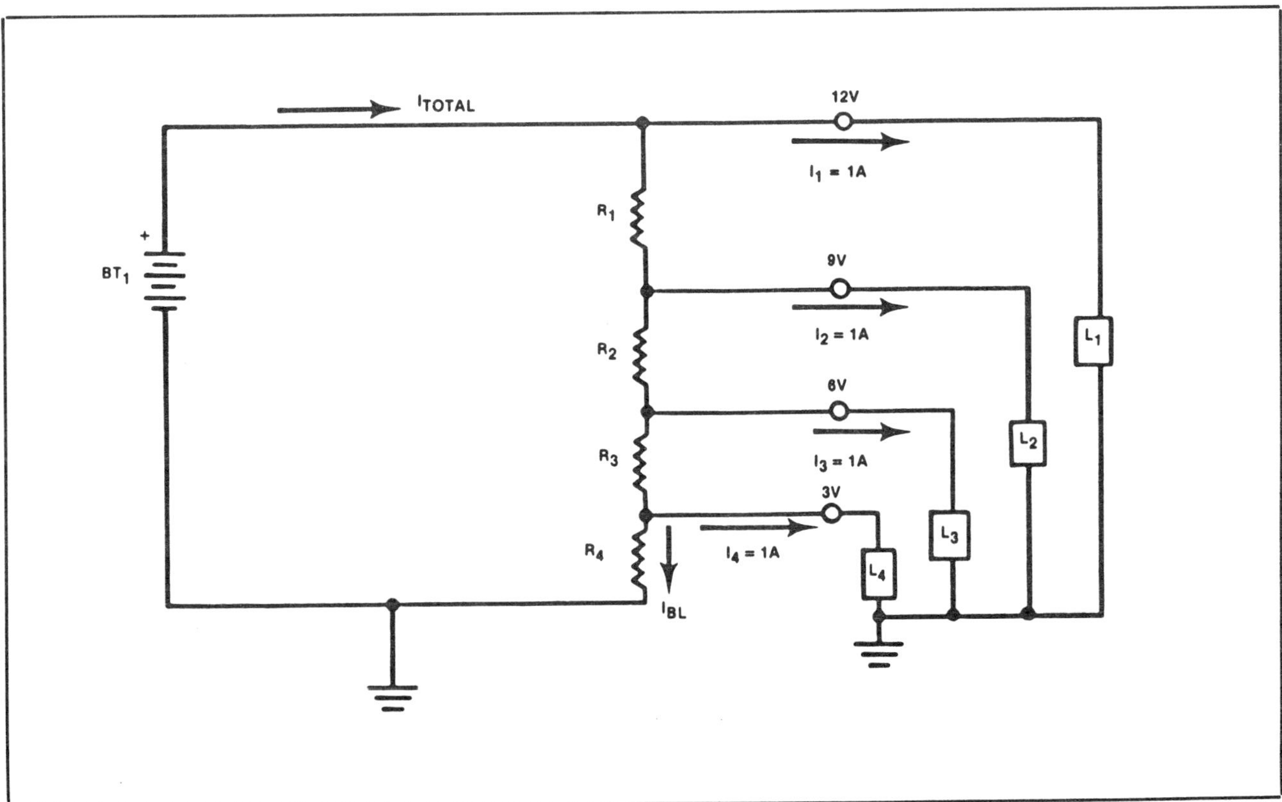

Fig. 7-7 A practical voltage divider. Unlike those shown in Section 6, this circuit is designed to furnish current to an external circuit.

resistor through which the least current flows, R_4 in this case. If we determine a convenient value for this divider resistor, the other resistances in the divider may be calculated simply.

There are only *two* real limits on the value chosen for this resistor. First, it cannot be so low as to cause a current flow that would strain the power source. Second, its resistance must not be so great that when used with the other resistors in the divider whose values are dependent on it, the current required by the loads cannot be driven through the first divider resistor. In most designs involving dividers which must deliver substantial currents, the bleeder resistor is chosen so that the current through it is about 10% of the total current furnished by the source. Once this current is determined, the values of all the resistances in the divider may be calculated.

Let us consider the voltage divider and loads shown in Fig. 7-7. The total current delivered by the battery, BT_1, is the sum of the bleeder resistor current and the currents delivered to each of the loads, or 4 A + $I_{bleeder}$. To obtain a bleeder current that is roughly 10% of the total current, we would choose a bleeder resistor that would pass a current of about 0.4 A for R_4. Since the potential difference across R_4 is 3 volts, the calculated value for R_4 is 7.5Ω.

The current flowing through R_3 is made up of the bleeder current, 0.4 A, and the current to load L_4, 1 A, or 1.4 A in all. Since the voltage drop across R_3 is also 3 volts (not 6 volts, as one might guess without looking carefully), the resistance of R_3 will be 2.1Ω. The resistance of R_2 is 3 volts/$I_{BL} + I_3 + I_4$ or 1.25Ω, and the resistance of R_1 is given by 3 volts /$I_{BL} + I_2 + I_3 + I_4$ or 0.88Ω. Practically, a 0.88Ω resistor would be a bit difficult to obtain. Therefore, if required to build a divider providing the currents and potential differences shown in Fig. 7-7, it would be necessary to recalculate the divider resistor bringing R_1 closer to 1Ω in value.

This can be done by choosing a bleeder current of 0.1 A rather than the 0.4 A previously chosen. In this case, R_4 = 30Ω, R_3 = 2.7Ω, R_2 = 1.4Ω, and R_1 = 0.96Ω. A practical circuit may be built by using 5% tolerance resistors of 30, 2.7, 1.5, and 1Ω. The required power for these resistors is respectively: 0.3 watt, 3.2 watts, 6.6 watts, 9.6 watts. Standard units of 0.5 or 1 watt for R_4, 5 watts for R_3, 7.5 or 10 watts for R_2, and 12 or 15 watts for R_1 would be selected, depending on the reliability and overall operating temperature requirements. A carbon composition resistor could be used for the bleeder, but the higher dissipation units, R_1 through R_3, would have to be wirewound.

A. Self-test Questions:

1. Complete the following table:

First Color Band	Second Color Band	Third Color Band	Fourth Color Band	Nominal Resistance	Maximum Resistance	Minimum Resistance
orange	white	red	silver	____	____	____
____	____	____	silver	10 MΩ	____	____
____	____	____	red	____	693.6Ω	666.4Ω
brown	black	orange	gold	____	____	____
gray	red	gold	gold	____	____	____
____	____	____	silver	470Ω	____	____
brown	black	blue	silver	____	____	____
____	____	____	____	530 KΩ	556.5 KΩ	503.5 KΩ
____	____	____	red	430Ω	____	____
violet	green	red	gold	____	____	____

2. Complete the following sentences.

a. Resistor power dissipation may be determined from the _______ of the resistor for resistors with a power dissipation of 2 watts or less.

b. A 2-watt, 5%, 100Ω resistor will dissipate *more/less* heat than a 0.1 watt, 10%, 1000 ohm resistor.

c. For high power dissipation *wire-wound/carbon composition/film* resistors are most generally used.

d. _______ and _______ are two of the principal drawbacks in the use of wire-wound resistors.

e. The most frequently found components in avionics circuits are __________ resistors.

f. The resistance of a carbon composition resistor will *slowly change/remain constant* throughout its life span.

g. Carbon composition resistors must be treated carefully; they are sensitive to:

a. ____________________

b. ____________________

c. ____________________

h. In high voltage or low-noise applications _______ or _______ resistors should be used, _______ resistors should be avoided.

i. A shunt resistor is placed in *series/parallel* with its load; it is primarily a *voltage dropping/current carrying* device.

j. A series resistor may be used to _______ voltage or limit _______.

k. Film resistors are available in power dissipation ratings from _______ watt to _______ watts.

l. Carbon composition resistors *may/may not* be used to replace film or wire-wound types.

m. A burned-out, 10% tolerance resistor *may/may not* be replaced by a 5% tolerance resistor of equal resistance, wattage, and reliability.

n. Connecting two 10% tolerance resistors in parallel *will/will not* provide the same precision as one 5% tolerance resistor.

o. A bleeder resistor is used in a _______ circuit; it passes *more/less* current than any other resistor in that circuit.

p. Two 100Ω, 1-watt resistors in parallel *may/may not* be used where a 50Ω 2-watt load is required for test purposes.

q. Choice of a meter shunt resistance is based on the _______ and the _______ of the meter.

r. The choice of bleeder resistor current is based on _______ _______. The value may be adjusted to _______ _______.

s. If a shunt resistor burns up, creating an open circuit, the component with which it is in parallel may be destroyed by __________.

t. If a series voltage dropping resistor opens, the load in series with it will *probably/probably not* be destroyed.

u. The high-end range of an ammeter, that is, its ability to measure high current, can be increased by putting a _______ in _______ with the meter.

B. Questions for Thought:

1. A small navigation warning system requires 5 volts at 1 ampere and 12 volts at 200 mA. Design a voltage divider that will supply 5.5 volts at 1.2 amperes and 12 volts at 200 mA from an aircraft 27 volt DC supply (the additional 0.5 volts and 200 mA are for a regulating circuit that would have to be in-

cluded). Draw the schematic diagram and calculate the theoretical values of the divider resistors and their power dissipation ratings. Assume that 5% resistors will be used. Could a carbon composition resistor be used for any of the resistors in the divider?

2. It is desired to install a small DC motor and a buzzer in series in a small aircraft so that the buzzer will sound whenever current is furnished to the motor. The aircraft power supply furnishes 12 volts, the motor requires 3 A and has a resistance of 2Ω, the buzzer has a resistance of 6Ω and is rated at a current of 1 A. Draw the circuit. What additional component would be required?

3. As the owner of a small avionics repair shop you are faced with the problem of choosing a stock of resistors for replacement purposes. What decisions would you be required to make and what information would you use to make these decisions?

4. Outline the steps one could follow in the design of

   a. a series voltage dropping circuit

   b. a meter shunt

   c. a voltage divider

C. *Experiments and Demonstrations:*

a. Make the test set-up shown in Fig. 7-8. The digital multimeter should be set to measure current.

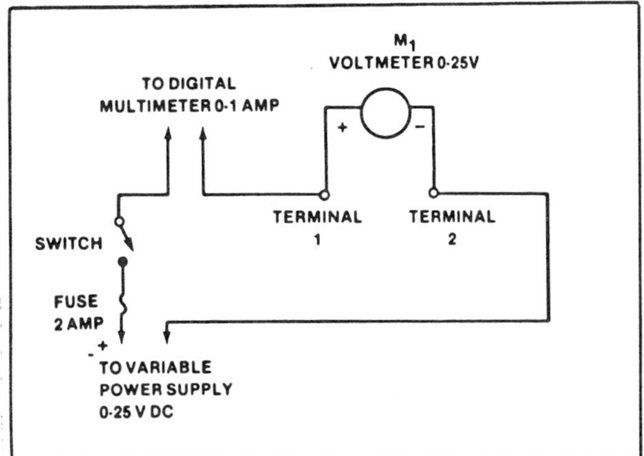

Fig. 7-8

b. With the power supply set at 1 volt, connect a 100Ω, 1/4 watt, 10% carbon composition resistor between terminals 1 and 2 of the test set-up. Record the current and voltage as indicated on the digital multimeter and voltmeter M_1. Using an electronic temperature probe or quick-acting thermometer, measure and record the temperature of the resistor. Calculate the actual resistance of the resistor. Draw up a table of your results in this form:

Voltage	Current	Calculated Resistance	Temperature

c. Increase the voltage of the power supply so that M_1 reading increases by 0.5 volt. Allow a minute or two for the temperature of the resistor to stabilize and record the voltage, current, and temperature. Calculate the resistance of the resistor.

d. Repeat step c until the resistor discolors and smokes (a voltage of 10 to 15 volts is normally sufficient). Carefully record the current, temperature, and voltage. Open the switch to shut off the current and reduce the power supply voltage to 0 volts.

e. Graph the calculated resistance of the 100Ω resistor against its temperature. Is a carbon composition resistor linear near the temperature at which it begins to smoke?

f. Graph the calculated resistance of the 100Ω resistor against the power being dissipated by it. How much power will the resistor dissipate before it begins to change its value? (Note: this depends on the temperature of the air around the resistor.)

g. Allow the resistor to cool. Measure its resistance using an ohmmeter. Is the resistor still within its marked tolerance?

2. Repeat experiment 1 using a 100Ω, 1/8-watt, 10% resistor, varying the power supply output so that each step represents a 20 mA increase instead of the 0.5 volt steps used in experiment 1. *Caution:* the resistor will burn up before the experiment is completed. Compare the shape of the graphs obtained with those obtained in experiment 1. Does the 1/8-watt resistor show as much

relative tolerance to an overload as the 1/4-watt resistor?

3. Repeat experiments 1 and 2 while heating the test resistors with a 25-watt soldering iron held beneath the resistor and about 2 inches away from the resistor body. The heated air rising from the soldering iron should be allowed to heat the resistor for a couple of minutes before starting the experiment and during the experiment itself. Compare the results with those obtained in experiments 1 and 2. If we define power dissipation of a resistor as the maximum power the resistor may dissipate without affecting its nominal resistance, what are the new *power dissipation* ratings of the heated resistors?

SECTION VIII

Variable Resistors

This section contains material on the construction and practical application of electronic components whose theory of operation has been covered in earlier sections of this book. A number of components, such as capacitors, transformers, and coils of various kinds, will be discussed from both a theoretical and a practical point of view in the next book in this series. That book will deal with the more complex aspects of DC circuits, magnetic effects, and with that period of time before a DC circuit has settled down to its steady-state condition.

After finishing this section, you will be familiar with those basic DC circuits which show a linear relationship between voltage and current. One non-linear component, the *thermistor*, will be introduced in this section, since it is actually a special type of resistor.

A. Potentiometers, Trimmers, & Variable Resistors

Potentiometers, trimmers, and *variable resistors* are actually three different names for the same general type of component. This component is a resistor whose resistance value can be varied within a certain range by moving a contact point to a different place along a resistance element. The last term in the paragraph title, "variable resistor," is the most widely accepted term, and can be used to refer to a *two-terminal device* or a *three-terminal device.* A potentiometer, on the other hand, is always a *three-terminal device.* In the case of variable resistors, a two-terminal device is a component with two connection points. The resistance between them may be varied, by either rotating a shaft, turning a screw, or otherwise adjusting the position of a movable contact point. A three-terminal potentiometer consists of two terminals between which there is a fixed resistance element and a moving contact which may be adjusted to any position along the fixed resistance element. Fig. 8-1 shows the schematic diagram symbols used for variable resistors.

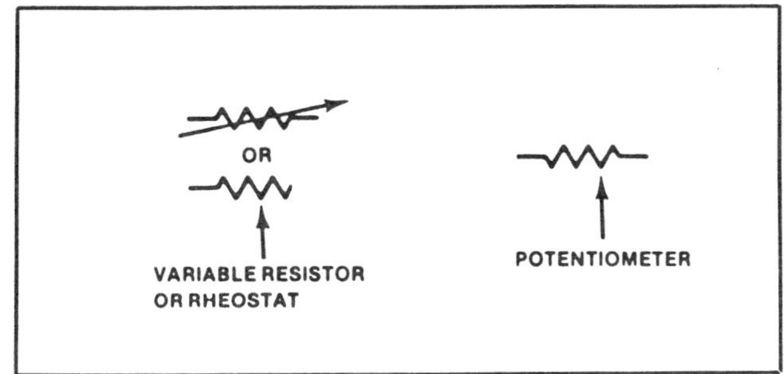

Fig. 8-1 Standard schematic symbols used on almost all manufacturers schematics.

All variable resistors used in industry today are constructed along the same basic principles, although their shape and the method of adjusting their resistance may vary. In all cases the variable resistor is composed of a resistance element (a coiled resistance wire, or a carbon or metallic film), and some sort of sliding contact arm which makes contact with the resistance element. Fig. 8-2 pictures one type of variable resistor, similar to that constructed in the Experiments and Demonstrations part of Section 5.

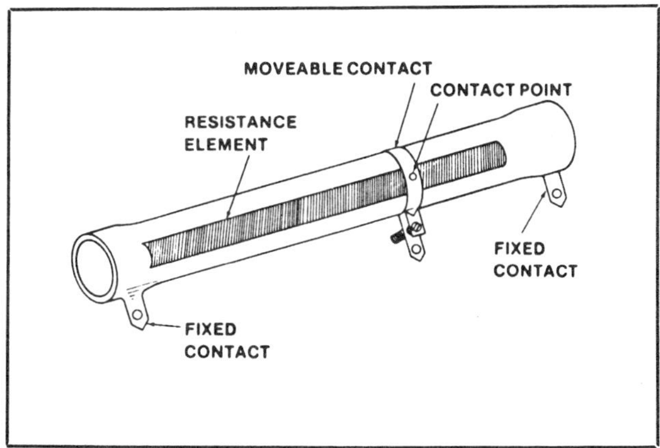

Fig. 8-2 A typical high power variable resistor with sliding contact.

Commercial units of this type are actually wirewound resistors. A length of resistance wire is wound on a ceramic core, and the wire ends are connected to terminals. The whole unit, except for the contact area, is then coated with an insulating glaze. Contact with the wire resistance element is made by a contact point on a sliding ring. This sliding contact is connected to the third terminal of the component. When the sliding ring is closest to one of the terminals, the value of the resistance between that terminal and the sliding contact is at a minimum. Moving the contact away from that terminal increases the resistance between the terminal and the slider. This is because there is a longer resistance path between them. Naturally, moving the slider away from one terminal brings it closer to the other terminal, thereby reducing the resistance between that terminal and the sliding contact.

Three-terminal variable resistors of this type are commercially available in resistances from 1Ω to 10 KΩ or more. Power dissipation ratings vary from 3 watts to 250 watts. The lower power units, from 3 to 15 watts, are those most frequently found in avionics equipment.

Another sort of variable resistor is the *rheostat*. This is a two-terminal device shown in Fig. 8-3.

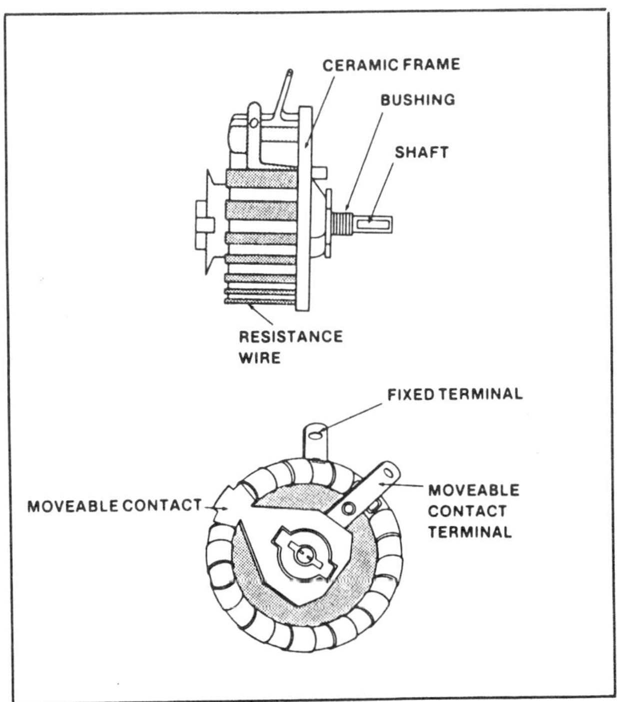

Fig. 8-3 A typical rheostat unit with a power dissipation of 25 watts.

In a rheostat, a horseshoe- or circular-shaped resistance element (either wirewound or covered with a carbon or metallic film resistance layer), is contacted by a sliding contact fixed to a shaft. A metal contact pad which has a very low resistance is used to connect one end of the resistance element to a terminal. As the shaft is rotated, the length of the resistance path between the metal contact pad and the sliding contact is varied. Thus, the resistance between the sliding contact and the terminal pad may be varied from nearly zero (when the sliding contact touches the metal pad) to the resistance of the entire resistance element (when the slider is at the end opposite the terminal pad). The sliding contact is connected to the second terminal of the rheostat.

Two-terminal rheostats are less common in avionics equipment than in power distribution or industrial equipment. Since the power dissipated in avionics circuits tends to be small, three-terminal components such as 2- or 4-watt potentiometers are usually substituted. Two-terminal rheostats are generally large, wirewound units built on an open ceramic frame for better cooling of the resistive element. Typically, the maximum resistance is low, from 1Ω to 100Ω, and the power dissipation rating is moderate, to 50 watts, although higher power units are available.

If we add a second metal pad and terminal to the other end of a rheostat resistance element, the result is a three-terminal component. This is the familar volume control or *potentiometer* pictured in Fig. 8-4. Unlike the two-terminal rheostat, the three-terminal potentiometer, commonly called a *pot*, is usually a low power dissipating component. Two watts is about the maximum for carbon or metal film devices, and four watts for wirewound types. Potentiometer resistance elements are manufactured in much higher resistances than rheostat elements. "Pots" with carbon elements of 50Ω up to 15 MΩ are available from distributors.

Most of the potentiometers and variable resistors used in avionics equipment are manufactured so that their resistance varies *linearly* along the element. That means the resistance between a fixed terminal and the sliding contact is exactly half the resistance of the element when the sliding contact is halfway along its track. On the other hand, most of the front-panel potentiometers used in entertainment equipment are not linear. This is because the human ear is not a linear

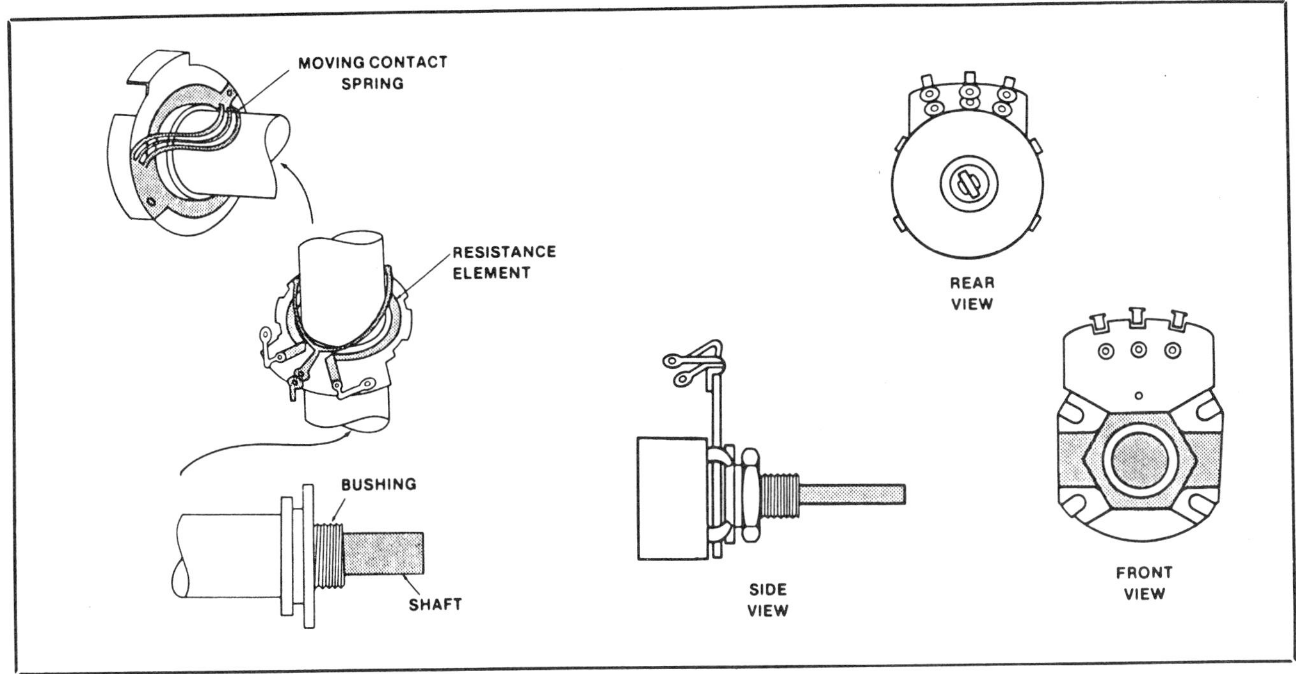

Fig. 8-4 A front-panel potentionmeter.

device. Using a linear potentiometer to control the volume of a radio receiver, alarm buzzer, or television set would result in too great a volume change for a given amount of shaft rotation at low volume levels. Too little change would occur at the higher volume levels.

For this reason, *logarithmic taper potentiometers* are used in volume control circuits. A simple experiment will demonstrate why these potentiometers are used. Let us connect an ohmmeter across the number 1 terminal and the number 2 (sliding contact) terminal of a logarithmic taper potentiometer whose control shaft is set to its extreme counterclockwise position. The ohmmeter will show non-linear increase in resistance like that shown in Fig. 8-5 as the control shaft is rotated to its extreme clockwise position. The non-linear resistance change seen here precisely compensates for the peculiarities of human hearing.

Since most of the potentiometers used in avionics equipment are not connected with volume control, the majority of the "pots" that the avionics technician will have to deal with will be linear. Note: Logarithmic or modified logarithmic (also called *audio-taper*) potentiometers should not be used where linear taper pots are required, and vice versa. Otherwise, it will be nearly impossible to set the control accurately.

Another type of variable resistor is the *trimmer potentiometer*. This is actually a family of types. Trimmers are small, usually three-terminal devices of varying shape, and are used for fine tuning adjustment of electronic circuits. They are built into critical avionics circuits which must be periodically adjusted for maximum performance. Trimmers used in industrial or entertainment electronics circuits resemble the resistance element and sliding contact of a small potentiometer, but do not have the casing or shaft. These units, pictured in Fig. 8-6, are usually adjusted with a screwdriver, or better, with a plastic alignment tool.

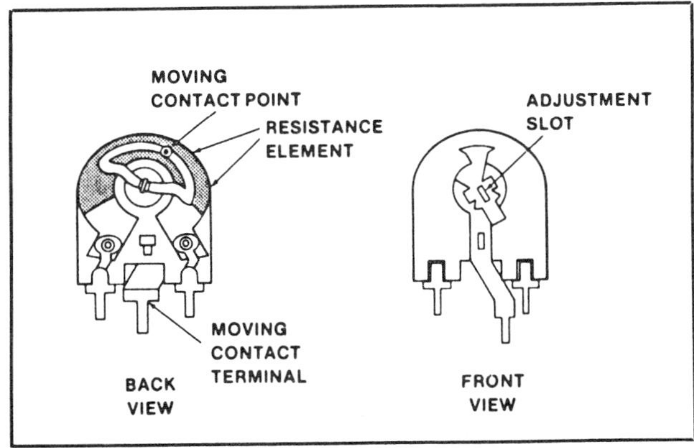

Fig. 8-6 Single turn trimmer potentionmeter — units of this type are more frequent in consumer equipment than in avionics applications.

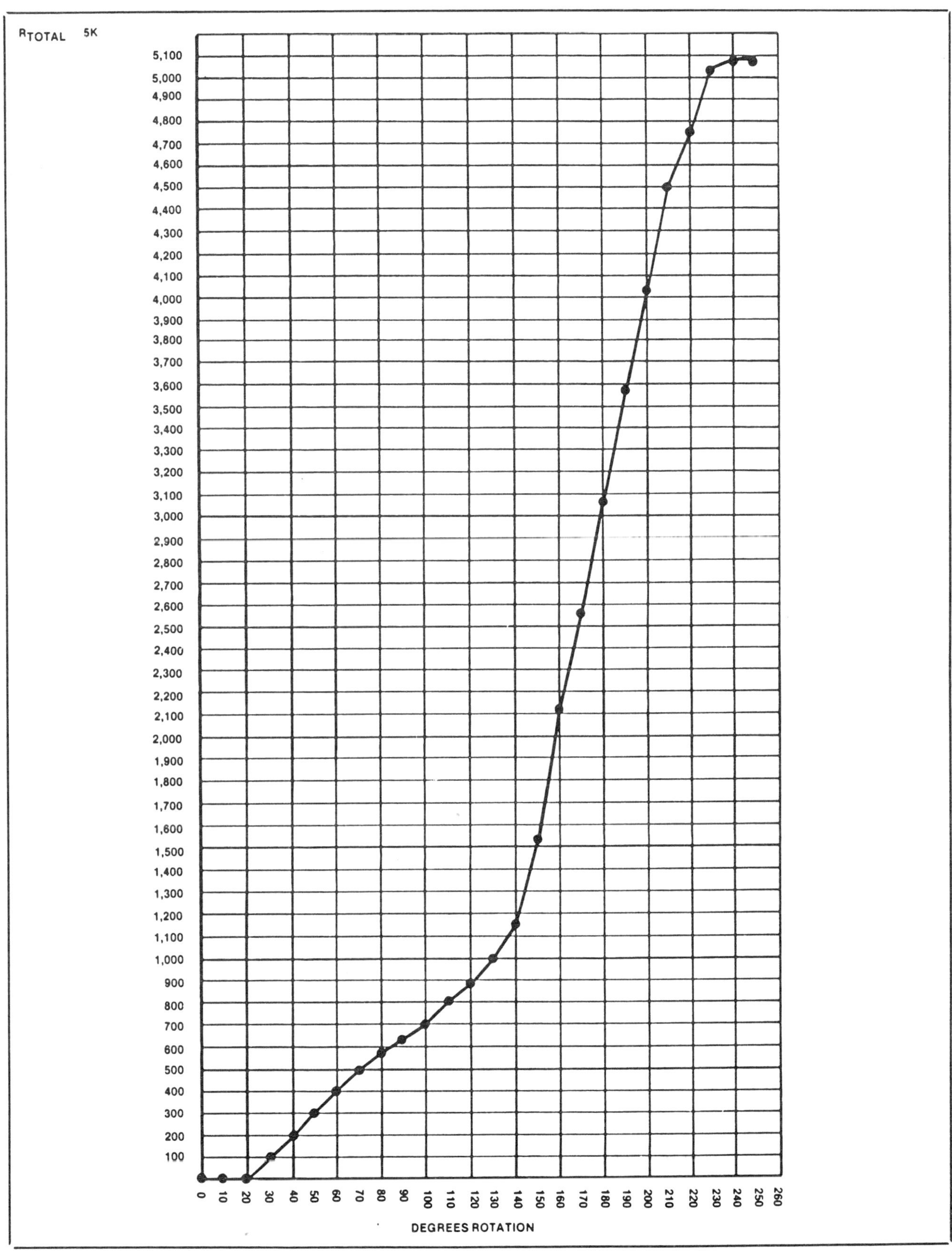

Fig. 8-5 A logarithmic or audio-taper potentiometer resistance curve. Notice the distortion of the curve due to the poor resolution of the particular unit measured.

Because of the small size of the resistance element and the relatively large contact area between the slider and the element, it is often difficult to set this type of trimmer exactly. The contact element, for example, may cover such a large area that even a minute rotation of the adjustment slot may cause a major change in the resistance between the sliding contact and the resistance element terminals. When this occurs, it is said that the *resolution*, or the ability to set the variable resistor to a specific resistance, is poor.

The fairly poor resolution obtained from the *single-turn trimmers* shown in Fig. 8-6 will frequently require use of a *multi-turn potentiometer* in avionics circuits. These are the square or rectangular trimmers pictured in Fig. 8-7.

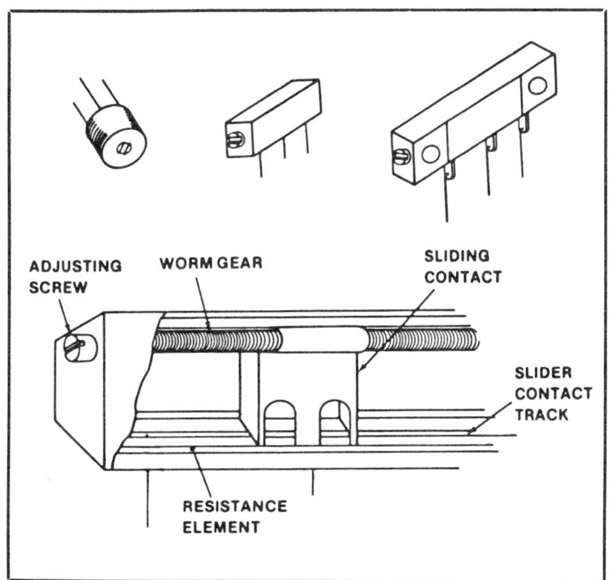

Fig. 8-7 Several forms of multi-turn trimmer potentiometers.

In these units, a screwdriver-adjusted worm gear moves a thin contact along the resistance element. Better resolution is obtained with these trimmers since the contact area of the sliding contact is smaller. Anywhere from four to twenty-five complete turns of the adjustment screw (depending on the trimmer model) are required to move the slider from one end of the resistance element to the other.

The mechanical linkage between the screwdriver-adjustable worm and the slider is open to a certain amount of mechanical inaccuracy. When we need extreme accuracy in setting a front-panel control to an exact setting, *multi-turn control potentiometers* (to be distinguished from multi-turn potentiometers) are used. Instead of a horseshoe-shaped resistance element, these potentiometers (Fig. 8-8), use a helix-shaped element. They often require ten full turns of the control shaft to move the sliding contact from one end of the resistance element helix to the other. Coupled with front-panel turn-counter knobs, these units are capable of extreme accuracy and may be reset to a pre-recorded position very precisely. In general, however, the high cost and large space required by these components limit their use.

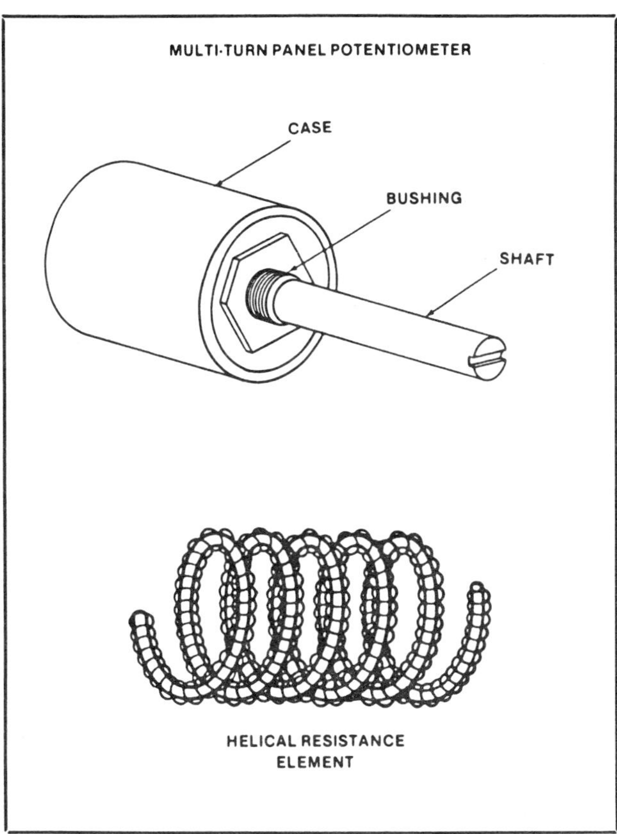

Fig. 8-8 Where great accuracy is required a multi-turn panel potentiometer is used.

1. *Variable resistor circuits*

a. *When to use a variable resistor*

The key to designing or understanding variable resistor circuits is the familiar principle of analyzing how electrical energy is controlled and applied to perform a specific job. If the purpose of a control is to vary the current *through* a circuit, a series-connected rheostat or a variable resistor in parallel with the load should be selected. On the other hand, if it is desired to vary the potential difference between a load input terminal and a ground reference point, a potentiometer connected as a voltage divider will be used.

The type of variable resistor selected will depend on how often an adjustment has to be made. If it is a question of changing a potentiometer setting fairly often, as for a communications receiver volume or gain control, a front-mounted unit is necessary. When the condition involves only infrequent re-tuning or *peaking*, (as when a unit is re-tuned to compensate for changed or aging components), multi-turn trimmers such as those pictured in Fig. 8-7 are employed. Large variable position sliding contact resistors pictured in Fig. 8-2 are sometimes used to adjust the output voltage of high-power standby generators, where it is usually not necessary to change the setting too often after initial tune-up and adjustment of the equipment. The following paragraphs offer a few practical examples of the many variable resistor applications found in the field of avionics.

b. Variable resistor voltage dividers

Because of the lack of standardization in power sources for ground equipment, additional components are often required to compensate for this variation. For example, single-phase mains power in different parts of the world may be 240, 220, or 120 volts, depending on the standard in force in a particular country. Consequently, the voltage available for a ground equipment site may vary considerably.

One way to compensate for this variation is to place a voltage divider in parallel with the load. A typical example of such a circuit is shown in Fig. 8-9, where the voltage divider is placed between the equipment power supply and the rest of the circuit.

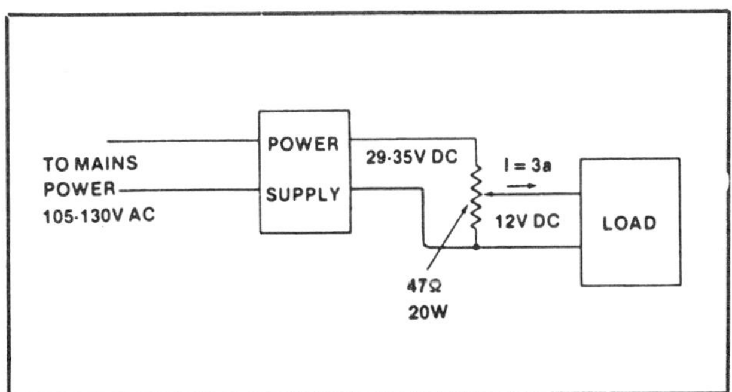

Fig. 8-9 A variable resistor voltage divider is frequently seen in power supply circuits.

To illustrate the application of a voltage divider, let us assume that the mains voltage varies between 105 and 130 volts AC, depending on the location of the site at which the equipment is to be installed. These extreme voltages will vary the power supplied to the ground equipment from 29 to 35 volts DC. Can a variable resistor voltage divider be designed so that the input voltage to the rest of the circuit can be adjusted to 12 volts DC at 3 A?

The solution to this problem is very much like calculating the resistance values of a standard voltage divider. With a voltage drop of 12 volts between the sliding contact and the terminal connected to ground, this section of the variable resistor should then have a resistance of 12/.3 or 40Ω. The top section of the voltage divider should have a voltage drop of 23 volts across it when the power supply is delivering 35 volts. Since the current through this section is the sum of the bleeder current and the current required by the other circuits in the unit, (3.3 A in all), the resistance of this section will have to be nearly 7Ω.

If we are in an area where the mains potential difference is only 105 volts, the power supply will deliver only 29 volts to the divider. In this case, the voltage drop across the top half of the divider will be 17 volts, and the resistance of this section should be slightly more than 5Ω. In either case, a 47 or 50Ω variable resistor may be used for the entire divider and the sliding contact can be adjusted to compensate for any mains voltage within the range of 105 to 130 volts AC. Calculation will show that a 20-watt resistor or a 25-watt unit should be chosen. This allows a safety factor.

c. A rheostat as a series lamp dimmer

Because of the varying light conditions in an aircraft cockpit, control of indicator intensities and dial illuminating lamps is necessary. A simple lamp dimming circuit consists of a rheostat in series with the lamp voltage supply line (the lamps themselves are in parallel, as shown in the schematic of Fig. 8-10). The purpose of this rheostat is to provide a 24-volt voltage drop so that the lamps will glow feebly when the rheostat is set at maximum resistance. When the rheostat is set near minimum resistance, the lamps will be at their rated current and brilliance.

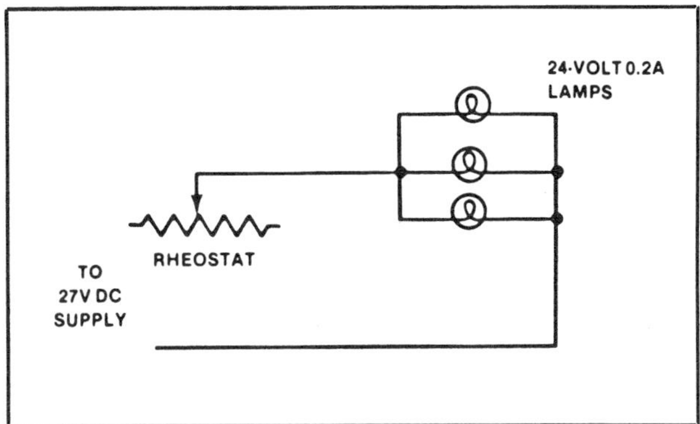

Fig. 8-10 A rheostat lamp dimmer, not a practical circuit, although it was used at times.

Since the total current required by all the lamps (Fig. 8-10) is 0.06 A, we can assume that this will be the usual circuit current under normal operating conditions. A 24-volt, 0.02 A lamp of the sort that would be used in this application, begins to glow when the voltage across it is in the neighborhood of 4.0 volts. Therefore, the maximum resistance of the rheostat must be high enough to provide a voltage drop of about 24 volts even when the current flow is slightly reduced. A unit with a 500Ω resistance element would be adequate. Since the most the unit is required to continuously dissipate is about 1.8 watts, a 3-or 5-watt rated component would be satisfactory.

d. Potentiometer applications

A frequent application of a panel-mounted potentiometer is as a load for one stage of a circuit. The potentiometer also serves as a means to select the voltage level input to the next stage. Fig. 8-11 shows a simplified schematic diagram of an integrated circuit amplifier whose input signal is derived across a section of potentiometer R_{50}. The 5 kΩ resistance element of the potentiometer acts as a load for a previous amplifier (not shown). A current from this previous stage is connected to one terminal of R_{50}, labelled E_{16}. Because of the current through the resistance element of R_{50} to ground, a potential drop is developed across the potentiometer resistance element. The sliding contact of the potentiometer, internally connected to the terminal marked E_{18}, may then be adjusted so that the potential difference between E_{18} and ground is some percentage of the potential difference across the entire resistance element.

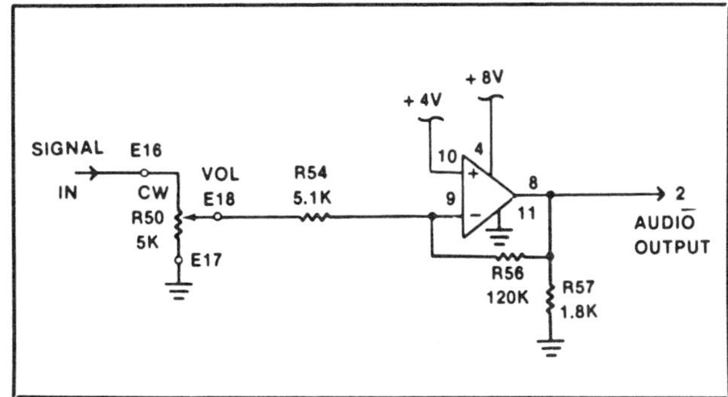

Fig. 8-11 A simplified schematic of a potentiometer volume control.

The *CW* notation near the E_{16} or "hot" terminal (the terminal with the highest potential when compared to ground) shows that clockwise rotation of the potentiometer shaft will bring the sliding contact closer to that terminal. The potential difference between the E_{18} terminal of the volume control and ground is applied to one of the input terminals of audio amplifier U_2C by input resistor R_{54}. Since the current through R_{54} is very small, there is very little voltage drop across it. The voltage present at pin 9 of U_2C will then primarily depend upon the potential difference between E_{16} and E_{17} (the ground end of the potentiometer) and the potentiometer shaft setting. In this particular application, the volume control for an intercom unit, an audio taper potentiometer would be selected.

B. Thermistors

In resistor design and production, much care is given to making these components as linear as possible. The object is to produce a resistor with a nearly constant resistance, regardless of temperature or current. This is especially important in avionics components, since airborne equipment must function over a wide range of temperature conditions without significant change in performance.

For certain applications, however, it is desirable to have a component whose resistance will vary appreciably and in a predictable way with a change in temperature. Such a component, a temperature-sensitive resistor, is called a *thermistor*. Fig. 8-12 pictures three widely-used forms of thermistor and also shows the schematic diagram symbol for this component.

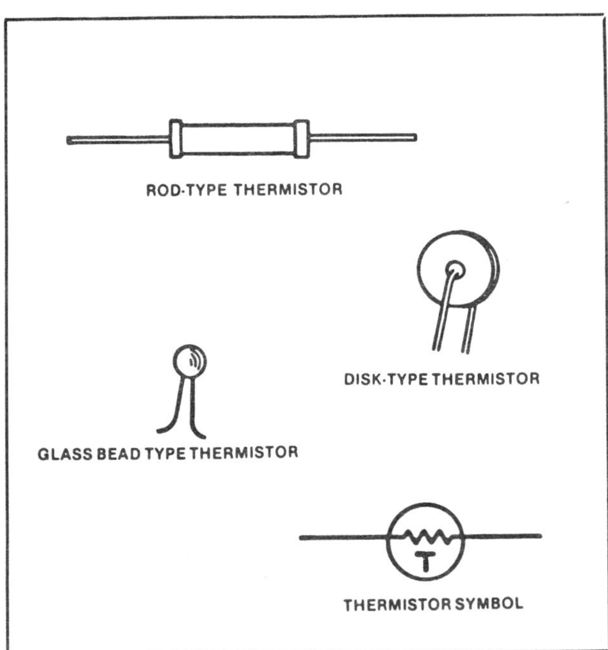

Fig. 8-12 The most commonly used thermistor types and symbols.

Although thermistors have been commercially available for nearly 50 years, the standardization that has brought so many benefits to the electronics industry has not yet been adopted for this component. So it is recommended that if it is necessary to replace a thermistor, a unit of the same type from the same manufacturer be used. Important thermistor characteristics include: direction of the coefficient of resistance change; the dissipation constant, symbolized by the Greek letter δ (delta); temperature coefficient of resistance, symbolized by the Greek letter α (alpha); the time constant, symbolized by the Greek letter τ (tau); and the initial resistance (R) at some fixed temperature.

The direction or sign of the coefficient of resistance change tells us whether the resistance of the thermistor will increase or decrease as the temperature of the unit increases. Nearly all thermistors in use today have a *negative* temperature coefficient. This means that the resistance of the thermistor will vary in a manner *opposite* to a change in the temperature of the unit: if the temperature of the thermistor *increases*, the resistance of the unit will *decrease*.

For thermistors, α or the temperature coefficient of resistance, is given in percent resistance change per degree centigrade. This is much higher than the temperature coefficient of resistance of metals such as copper. Although some special units may show a resistance decrease of as much as 300 times per degree centigrade, the thermistors normally found in avionics equipment possess an α of between -1.5 and -6 percent per degree centigrade, depending on the materials used and the design of the unit. The temperature coefficient of resistance, along with the other essential characteristics of a thermistor, can only be obtained from the manufacturer's specification sheets. They are almost never marked on the components.

The time constant, τ, of a thermistor tells us how long it takes the component to respond to a change in the temperature of the air around it. It is defined as the amount of time it takes for a thermistor to change its temperature from some initial temperature to 63% of the temperature of the air surrounding the unit. Average units found in avionics equipment have time constants ranging from about 10 to 50 seconds.

As we shall see in the next section, the time constant of a thermistor plays a role in determining the applications for which the component is suited. Incidentally, the figure 63% is not an arbitrary one. This number will recur when we discuss the charging of a capacitor and its storage of charge, which in some ways is analogous to the flow of heat energy in a system.

The nominal resistance of a thermistor is the actual resistance of the unit at a particular reference temperature. If the temperature coefficient of resistance is accurately known, the reference temperature chosen is not important. This is true, if it is within or near the range at which the component is to be operated. Room temperature, 20 °C, is a convenient reference point. Some manufacturers may use a different reference temperature, depending upon their testing techniques.

The dissipation constant, δ, is a critical characteristic which gives the ratio between the amount of current passing through a thermistor and the temperature increase due to the heating effect of this current. This characteristic is usually given in milliwatts per degree centigrade. Commercially available units range from a δ of about .5 to as much as 30 mw/°C. As we shall see in the following paragraph, these characteristics permit the thermistor to be used in a number of special applications that require a component with a non-linear voltage-current characteristic.

1. Thermistor applications

There are three primary circuit types in which thermistors may be used: temperature-sensing circuits, surge protection circuits, and compensating or regulating circuits. Only the last is used in avionics. This is because the long time constants involved make thermistors less suitable than other components for temperature-sensing or surge protection circuits in avionics equipment.

For temperature compensation, where it is necessary to counterbalance a change in the characteristics of another component due to a gradual rise in temperature, thermistors may prove quite useful. The simplified circuit shown in Fig. 8-13 shows a typical application. In this illustration, the diode resistance will increase as it gradually heats up during normal operation. Now, if it were necessary to keep the amount of current flowing through the circuit nearly constant, this could be accomplished by using a thermistor whose thermal coefficient of resistance was opposite to that of the diode, mounting the two components so that they will have roughly the same temperature. The total series resistance of the two components would then remain constant. As the components warm up, the diode resistance increases, and the thermistor resistance decreases.

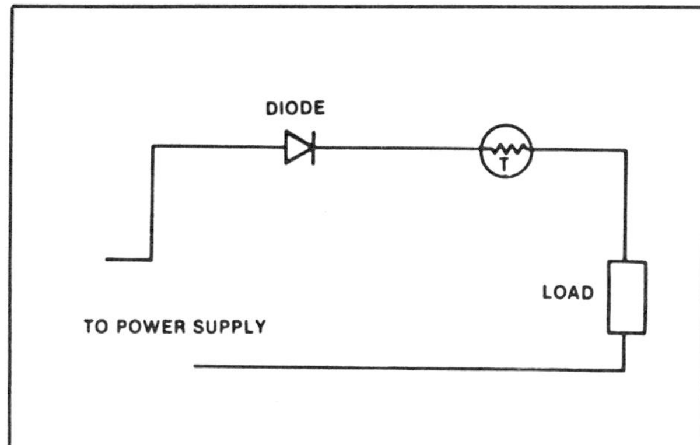

Fig. 8-13 Thermistor compensation for diode — one application of the thermistor.

Another application of thermistors sometimes seen in avionics equipment is the automatic output level control pictured in the simplified schematic of Fig. 8-14. The integrated circuit operational amplifier symbolized by the triangle is a linear device whose output may be adjusted to be a certain number of times greater than the input. Thus a weak input signal may be *amplified* by a factor of five, ten, or more.

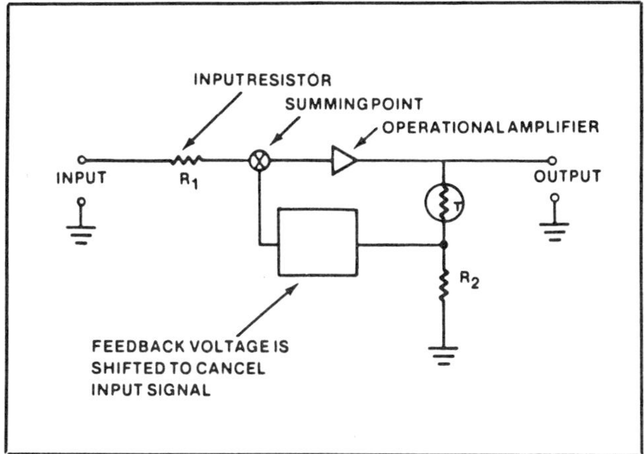

Fig. 8-14 Automatic output level control using thermistor.

In some cases, this sort of linear response is not desired. For example, if the input signal is much larger than normal, the amplified output signal could be great enough to destroy following components or even burn out the operational amplifier itself. One way to avoid this is to feed back a portion of the amplifier output signal to cancel some of the input signal to the operational amplifier. This effectively reduces the output of the amplifier.

Fig. 8-14 shows how a thermistor may be used for this sort of control. The thermistor forms part of a voltage divider between the amplifier output and ground. When the output level of the amplifier is low, and the resistance of the thermistor is high, the potential difference across resistor R_2 will be rather small. The feedback voltage, which is the same as the potential difference across R_2 will be small, cancelling very little of the input signal.

If the input signal to the amplifier increases, the output signal also increases, causing more current to flow through the thermistor. When this occurs, the additional heat generated by the increased flow has an effect on the thermistor, causing the resistance of the thermistor to decrease. The voltage drop across R_2 will be proportionally much greater. The larger feedback signal will cancel a greater portion of the input signal. Therefore, the output of the amplifier will drop. In fact, through correct selection of R_2 and the thermistor, we could even cause the input signal to the amplifier to be completely cancelled out when too large an input is received from previous circuits. This method of overload protection is very useful in a number of applications.

A. Self-test Questions:

1. Complete the following sentences:

   a. A thermistor possesses a *negative/ positive* temperature coefficient of resistance. Its resistance __________ as its temperature increases.

   b. The *resolution* of a variable resistor is a measure of the ability __________.

   c. The familiar front-panel *volume* control is most frequently a __________-__________.

   d. Trimmer potentiometers are frequently used for __________.

   e. A rheostat is a __________-terminal device; a potentiometer is a __________-terminal device.

   f. The power dissipation rating of most of the potentiometers used in avionics equipment is less than __________ watts.

   g. Carbon potentiometers are generally available with total resistances of ______Ω to ______Ω.

   h. The time constant τ (tau) of a thermistor is defined as __________.

2. Connect the thermistor characteristics with their respective symbols:

δ	dissipation constant
α	temperature coefficient of resistance
τ	time constant
R_{20}	nominal resistance

B. Questions for Thought:

1. It is necessary to connect a 6.3-volt, 0.3 A indicator bulb across a 12-volt DC aircraft generator in such a way that its light intensity can be controlled, but it is also necessary to make sure that the potential difference across the bulb never exceeds about 8 volts. The bulb begins to glow when the potential difference across it is about 2 volts.

   a. What *two* components could be used together to meet all these requirements?

   b. Draw the schematic of a circuit that would meet these requirements.

   c. Calculate the values of the components that would have to be used in your circuit.

2. Consider the circuit of Fig. 8-14. In what way would this circuit function differently if the thermistor were replaced by a resistor? Would there still be some automatic control over the amplifier output?

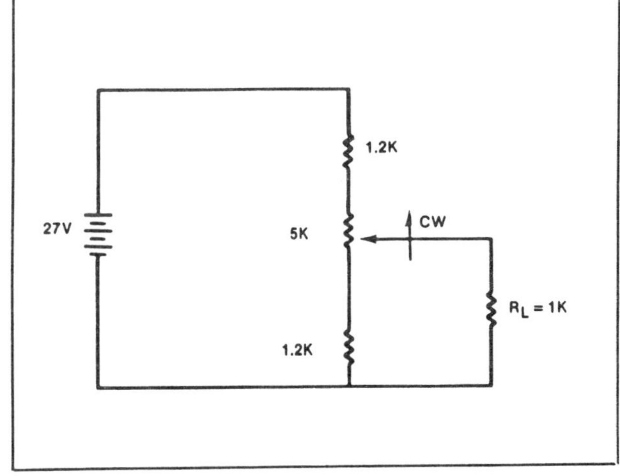

Fig. 8-15

3. Given the circuit of Fig. 8-15:

   a. Calculate the voltage across R_L and the current through it when the potentiometer is in its full clockwise (CW) position.

   b. Calculate the voltage across R_L and the current through it when the potentio-

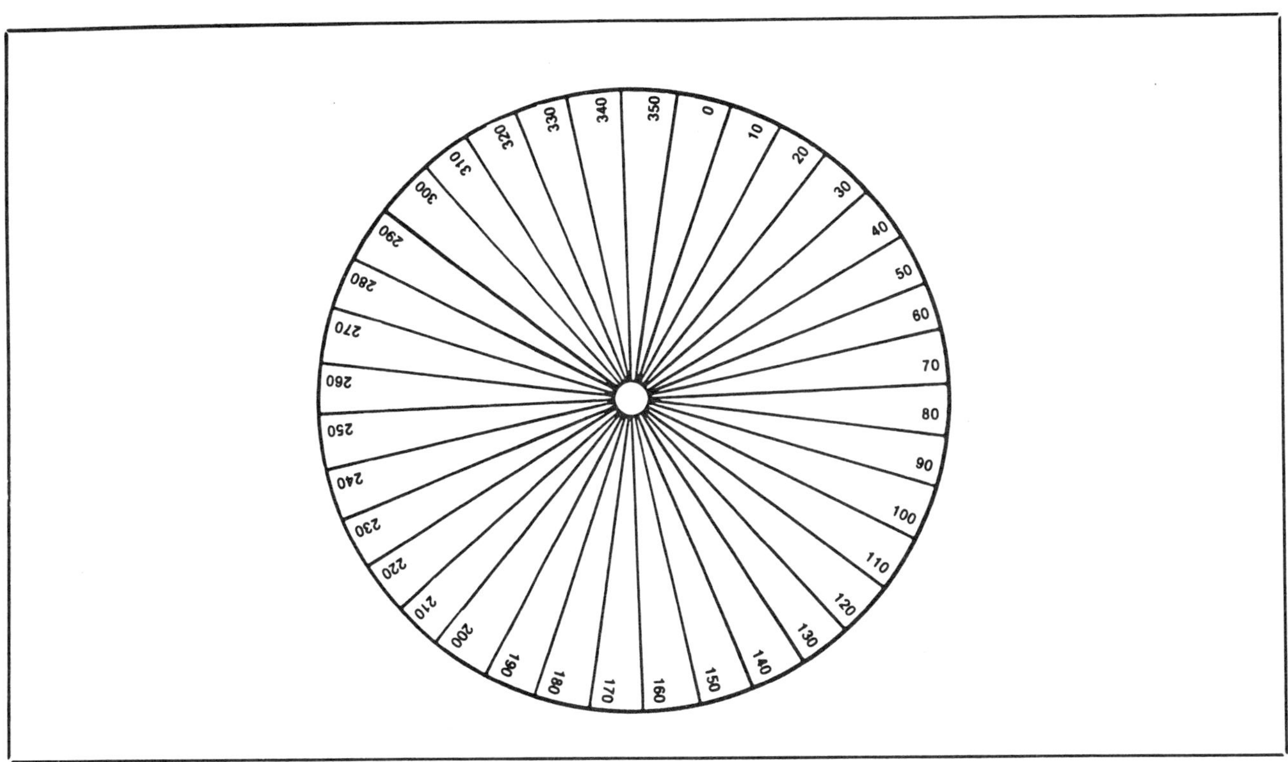

Fig. 8-16 Sample cardboard protractor.

meter is in its full counterclockwise (CCW) position.

c. For what practical purpose might a circuit like this be used? What are the probable functions of the two 1.2 kΩ resistors?

4. Refer to the rheostat lamp dimmer example of Fig. 8-10. What would happen if one or two of the lamps burned out? Would we still retain full brightness control over the remaining lamp?

   a. Could this circuit be modified to permit full control over the brightness of the remaining lamp, even if two of the lamps burned out?

5. Design an experiment that will allow you to determine all the significant characteristics of an unmarked thermistor using only a digital multimeter, a variable DC power supply, and a wristwatch with a sweep second hand. (Hint: assume that room temperature is 20 °C and that the human body temperature is 37 °C.)

C. Experiments and Demonstrations:

1. a. Make a stiff cardboard protractor like that pictured in Fig. 8-16 and install it on the bushing of a 5 kΩ linear taper carbon potentiomenter along with a pointer knob fixed to the potentiometer shaft. Adjust the position of the protractor so that the pointer knob will coincide with 0° when the shaft is in its extreme CCW position. Tighten the bushing nut to hold the protractor firmly in place.

   b. Turn the pointer knob to its extreme CW position. Through how many degrees did the knob turn? If this rotation represents a resistance change of 5 kΩ, how many ohms are represented by a 1° rotation?

   c. Return the shaft to its extreme CCW position. Connect a digital multimeter to terminals 1 and 2 of the potentiometer. Adjust the multimeter to an appropriate "ohms" scale and measure the resistance between the terminals as the potentiometer shaft is rotated in 10° increments. Record your results in a table of this form:

Pointer Settings (Degrees)	Resistance	Change in Resistance from Last Setting
0°		
10°		

d. Are the changes in resistance nearly equal for equal 10° increments of the shaft setting?

e. Set the multimeter to read 10 volts DC, and connect a 10 volt DC source between terminals 1 and 3 of the potentiometer. Starting at 0°, record the voltage reading for 10° increments of the shaft position. Throughtout the range of the potentiometer are the voltage increments equal? Can you set your voltage divider for a reading of 5.65 volts, exactly?

2. Repeat experiment 1 using a 5 kΩ audio taper potentiometer. Note the differences in linearity between the two potentiometers. Which potentiometer gives better control and resolution over the first 500 ohms? Over the last 500 ohms?

3. Using a number of bead, rod, and disk type thermistors, perform the experiment designed in part B, question 5.

SECTION IX

Chemical Cells

In this final section, we will discuss chemical cells which are increasingly being used in avionics as power sources for portable equipment as well as standby sources for power-line operated units.

Before the days of extensive power networks, many isolated installations, farmhouses, and equipment sites used a combination of gasoline-or wind-powered generators and lead-acid storage batteries for electric power. With the spread of dependable power line networks, chemical cells were more or less relegated to providing portable power for starting automobiles, light aircraft, for portable radios, etc. Recently, with the increased use of computer techniques in aircraft navigation, and with the appearance of ground support equipment located in inaccessible areas, the need for efficient chemical cells for *both* principal and standby service has become more important. For example, in the latest instrument landing system, much of the signal control and monitoring will utilize computer-type electronic memory. Consequently, even a temporary loss of power, such as experienced during the flickering of electric lights during a thunderstorm, would be sufficient to erase this memory. To guard against this, uninterruptible power supplies using chemical cells are built into the system.

Within the next few years, many of the new cells being developed will play an increasing role in the field of avionics. For this reason, the avionics technician should be familiar with these units. In this section, each of the cell types presently in use, or expected to be used in avionic applications, will be discussed. The features and drawbacks of each type of cell will be analyzed, with an eye toward providing the future technician with a good background for understanding space-age developments in this area.

A. Chemical Cells and Voltage Rises in a Circuit

In addition to the basic theory of the chemical cell, discussed in Section 2, a practical understanding of the characteristics of modern cells requires a review of how a cell functions in a circuit. The chemical cell, as you will recall, provides an electric field and potential difference between its positive and negative poles. The cell does this by producing an excess of electrons at its negative pole and an excess of positive ions (equivalent to a deficiency of electrons), at its positive pole. If a closed conductive path is connected between the poles of a cell, electrons will begin to drift along this path or circuit.

Even though the charge carriers are electrons, we use the conventional notion of current flow in the direction *opposite* to that of the electron drift. Therefore, we can say that the current flow through a circuit is from the positive pole to the negative pole of the source. In simple DC circuits, the electron flow and current flow concepts may be interchanged without any confusion. This is because the mathematical expressions used to describe current flow and electron drift are basically the same.

Let us examine a simple circuit. In passing through the loads in a circuit, a current causes energy to be converted from electrical energy to some other form, such as heat. The energy converted by the loads in the circuit must be supplied by the source. Putting it in a more formal way:

Energy used by loads in a circuit =
Energy supplied by source(s) in the circuit

Equation 9-1

Equation 9-1 may be rewritten to give us:

Energy supplied by sources − Energy used = 0

Equation 9-2

In the sense of equation 9-2, it is convenient to speak of a power source (e.g., chemical cell) as an *energy or voltage rise* in a circuit, and a load or resistance as an *energy or voltage drop*. This allows us to state the idea of equation 9-2 in an easy-to-remember formula:

$$V_{drop} - V_{rise} = 0$$

Equation 9-3

This equation means that the voltage drops in a circuit are exactly equal to the voltage rises. That is, all the energy used in a circuit is supplied by the sources, and all the energy supplied by the sources is used up in the circuit.

In a circuit such as that shown in Fig. 9-1, the voltage rises are *exactly equal* to the voltage drops. This is easily seen if we use a method of adding voltage rises and voltage drops used in more advanced DC and AC circuit analysis. This method is to begin at any point in the schematic diagram of the circuit and trace the current flow through the circuit and back to that point, taking all branches into account. As we trace the current flow, we can record every voltage drop as a negative number ($V_{drop} = I \times R$) and every voltage rise as a positive number. Thus, a voltage rise will occur when we pass through a voltage source *in the direction of the conventional current* (from its negative pole to its positive pole).

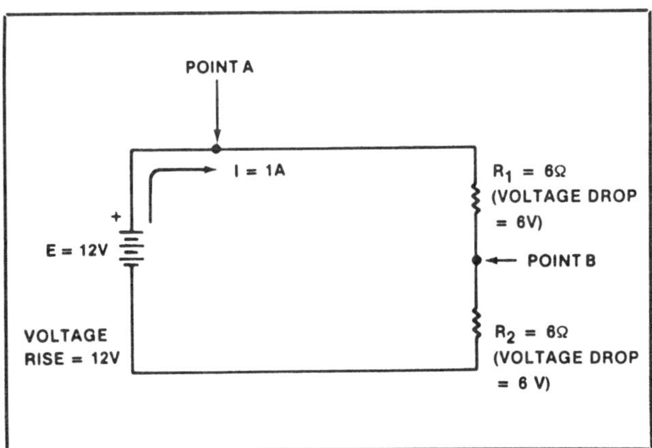

Fig. 9-1 Adding the voltage rises and drops around a circuit.

Choose some arbitrary starting point in the schematic Fig. 9-1. Call our starting point *A*. We pass along the connecting wire in the direction of the current flow to the load represented by R_1. (Note that the wire itself is assumed to have no resistance for the purposes of this exercise.) The voltage drop across R_1, by Ohm's law, is equal to $I \times R$, or 6 volts. Therefore, the voltage drop between points A and B should be recorded as −6 volts. The voltage drop across load R_2, following along in the direction of the current flow, is also −6 volts. Continuing, we enter the source at its negative pole and leave at its positive pole, thus experiencing a voltage rise of +12 volts. This brings us back to point A. The sum of the voltage rises and drops recorded is:

−6 volts	drop across R_1
−6 volts	drop across R_2
+12 volts	rise across BT_1
0 volts	SUM

The sum of the voltage rises and drops in the circuit is zero, just as we saw in equation 9-3.

The total power used up in the two loads shown in Fig. 9-1 is $I^2R_1 + I^2R_2$ or 12 watts, while the total power furnished by the battery is $E_{battery} \times I_{circuit} = 12$ watts. We see that this circuit is a "closed" system. This means that all the energy supplied by the battery is used up in the circuit, and the power dissipated in the circuit is exactly the amount of power furnished by the battery.

At this point you are perhaps wondering why so much emphasis was placed on the direction in which our current flow tracing passed through the battery. The reason for this becomes evident in the schematic pictured in Fig. 9-2.

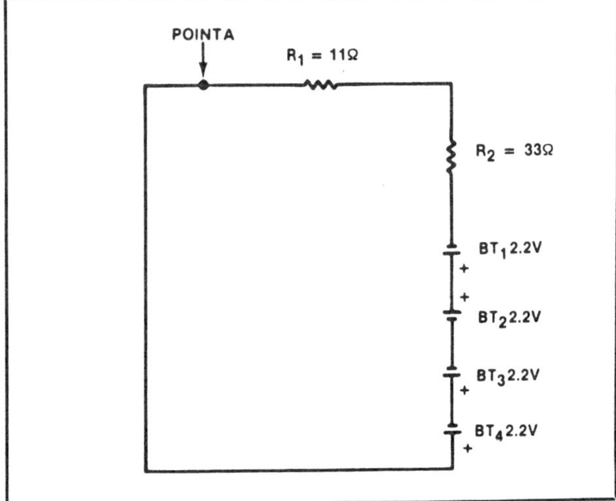

Fig. 9-2 Voltage rises and drops in a multi-cell circuit. Notice that one cell is connected in series bucking.

Here, we have a battery composed of four chemical cells, but one of the cells has been connected in series bucking. If we record the voltage rises and voltage drops in this circuit, the backward-connected cell in the series string must be recorded as a voltage drop, since we enter at the positive pole and leave at the negative pole. In later books in this series, we shall see that numerous equivalent circuits assume this form, showing sources connected in a bucking arrangement. We must be careful to notice the direction in which the total circuit current passes through these sources.

In Fig. 9-2, starting at point A, the voltage rises and voltage drops around the circuit are:

voltage drop due to R_1	-1.1 volts
voltage drop due to R_2	-3.3 volts
voltage rise due to BT_1	$+2.2$ volts
voltage drop due to BT_2	-2.2 volts
voltage rise due to BT_3	$+2.2$ volts
voltage rise due to BT_4	$+2.2$ volts
	0 volts

From the above discussion, we see that in a practical chemical cell, or battery composed of chemical cells, as in a circuit comprising any sort of DC sources, the voltage rises due to the sources must be *equal* to the voltage drops produced by the loads in the circuit.

In a new chemical cell, the voltage rise across the cell is determined by the type of material used for the positive and negative poles and by the electrolyte that permits ion and/or electron transfer between the poles. The *chemical system* of the cell is defined by its pole materials and electrolyte. Each chemical system has its own characteristic potential difference. This potential difference is independent of the structure or size of the cell. This characteristic potential difference, however, is generally somewhat higher than the *actual* potential difference across a cell measured by a voltmeter. When it is furnishing energy in a circuit, the potential difference across the poles of even a new chemical cell drops slightly. When a chemical cell has been used, or stored in an unused state for some time, the actual measured potential difference across its poles may differ significantly from the characteristic potential difference of its chemical system. The more a chemical cell is used (or the longer it is stored), the more its actual pole to pole potential difference will decrease. When the potential difference drops below the requirements of the circuit, the cell has to be replaced, or recharged, if it is a rechargeable unit.

B. Characteristics of Chemical Cells

Along with the chemical system and characteristic potential difference of a particular cell, there are a number of other cell characteristics with which the avionics technician should be familiar. First, all cells fall into one of two basic categories. They are either *primary cells*, which are not considered rechargeable, and are used until their potential difference drops below a certain value, or are *secondary cells*, which may be used and then recharged. Recharging is accomplished by forcing a *charging current* through the cell in the opposite direction of normal discharge. This reverses the chemical state within the cell, restoring it to its original condition.

Some sixty years after Alessandro Volta began to construct the chemical cells described in Section 2, two Frenchmen, working separately, developed early forms of the primary and secondary cells in widest use today. Gaston Plante in 1860 presented a paper to the French Academy of Science in which he described the construction of a cell using a chemical system composed of lead and lead dioxide electrodes and sulfuric acid and water as the electrolyte. The familiar, rechargeable lead-acid storage battery used in automobiles, aircraft, etc., is composed of secondary cells of this type.

At the same time, Georges Leclanche was at work developing the carbon-zinc primary cell which is named after him. Leclanche was trying to perfect a *dry* cell which would eliminate the hazards and inconveniences of a chemical system using a liquid electrolyte. In 1868 he was awarded a patent on a primary cell in which the electrolyte and the positive electrode were actually combined into a paste form. The carbon-zinc cells used in flashlights, transistor radios, etc. are, for the most part, only slight improvements of the Leclanche design.

A second bit of information the engineer, technician, or consumer needs to know about a chemical cell, is *how long* it will last in a particular circuit. That is, how much power it can provide before it must be recharged or thrown away. Unfortunately, this is by no means a simple question. For some chemical cells, the potential difference measured between the poles of the cell remains fairly constant for the usable lifetime of the cell. But after a period of time it decreases sharply as the cell nears its recharge or throw-away point. Within a very short time after this sort of cell begins its decline, its potential difference has dropped to the point where it will not power any circuit. Other sorts of chemical cells show a long, slow decline in their potential difference. The situation can be further complicated by the fact that some cells will recover a part of their original potential difference if allowed to rest for a period of time. How are we to determine when a cell whose nominal potential difference is 1.5 volts is depleted? When its potential difference falls to 1.2 volts, or 1 volt, or 0.5 volt? Obviously some sort of standard procedure for measuring cell capacity is required.

In general, chemical cell manufacturers have selected a so-called *end voltage* for each of their product lines. When the potential difference across the poles of a cell or battery reaches the end voltage, the unit must be replaced or recharged. For example, a standard carbon-zinc flashlight cell with an initial potential difference of 1.5 volts is said to have reached its end voltage when the potential difference drops to 0.8 volts.

Specifying an end voltage is only one step in the process. In order to determine how much power a cell may be expected to deliver, it is necessary to specify the amount and conditions of the current flow. Most manufacturers of chemical cells generally specify the amount of current flow by giving to each of their products an ampere-hour rating, abbreviated *a-h*. This figure is derived by connecting a new cell in a resistive circuit, and recording the current which should be kept nearly constant, the potential difference across the poles of the cell, and the time elapsed since the start of the test. When the potential difference falls to the end voltage value, the test is halted and the result tabulated. The current is multiplied by the length of time during which this current was furnished. For example, a large flashlight cell may have a rating of 6 a-h. Somewhere off in the fine print of the cell manufacturer's specification sheet, the conditions of this test are stated. In general, such a rating represents a current of 0.12 mA over a span of 50 hours.

The a-h rating of a chemical cell can be somewhat deceptive if we do not bear in mind the 50-hour discharge rate. For instance, a cell that will deliver a current of 120 mA for 50 hours will probably *not* deliver a 600 mA current for 10 hours, even though the a-h rating of both is 6 a-h. In fact, at a current flow of 600 mA, the flashlight cell may drop to its end voltage after two hours of service. This would bring the 2-hour discharge-rate service capacity down to I × time = 1.2 a-h, which is a service capacity considerably less than the 6 a-h at the 50-hour discharge rate. With very small circuit currents, in other words, at long discharge rates (especially if the cell is allowed to rest between periods of use), the service capacity may be even greater than the nominal 6 a-h at a 50-hour discharge rate.

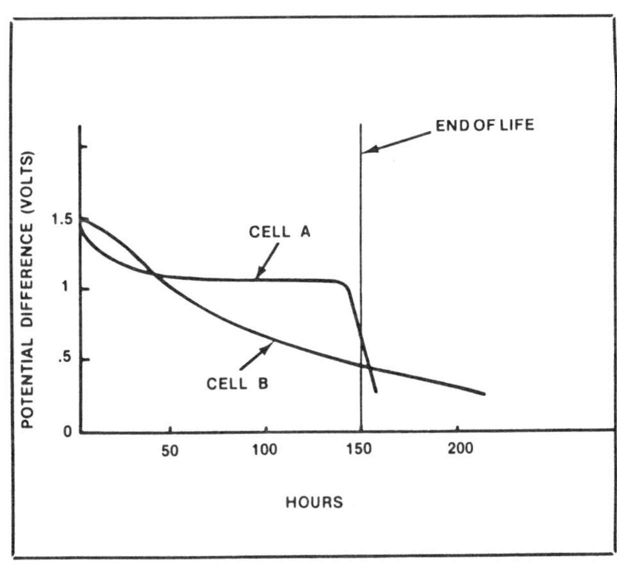

Fig. 9-3 Discharge curves for flat and sloping discharge types. Cell A may be a mercury cell and cell B a Leclanche cell.

Another factor related to the service capacity of a chemical cell, which has already been mentioned, is the shape of the discharge curve. In Fig. 9-3, the discharge curves for two types of primary cells are shown. Although the end voltage value for the two cells is the same, and the service capacity of both is about equal, cell *A* has a much flatter discharge curve than cell *B*. Practically speaking, this means a circuit which is sensitive to changes in its input voltage will have to be adjusted periodically if a type *B* cell is used. The ohmmeter section of most multimeters uses type B cells, and, therefore, it is necessary to adjust or "zero" the meter periodically. On the other hand, the exposure meter built into many modern cameras uses a type *A* cell, making readjustment unnecessary. When the cell reaches the end of its life, the unit simply ceases to function.

In addition to the service capacity and the shape of the discharge curve, it is important to know the effects of temperature upon the performance of a chemical cell. The most commonly used cells are quite sensitive to changes in temperature, although, most manufacturer's service capacity ratings are based on an environmental temperature (generally called the *ambient temperature*) of 20 °C. This is standard room temperature. Ordinary carbon-zinc flashlight cells are practically useless at temperatures in the neighborhood of −20 °C. The nickel-cadmium and lead-acid cells used in aircraft engine starting batteries also show a significant decline in service capacity at these temperatures. Unfortunately for the avionics technician who must be able to judge the service capacity of both airborne and ground equipment chemical cells, manufacturer's specifications can be rather foggy in this particular area. In fact, an authoritative engineering design journal has found temperature data in chemical cell and battery catalogs, with a few exceptions, to be incomplete and imprecise.

Shelf life of chemical cells is directly related to temperature. This characteristic is of extreme importance to the avionics technician, since all primary cell and some secondary cell backup battery systems essentially put the cells "on the shelf". This means the cells are disconnected from the equipment until a failure of the commercial power line or prime power source brings them into service. Under this condition, an electrochemical process within the cell causes a slow but continuous discharge in many types of chemical cells. This process is called *local action* and is based on the non-uniformity of the cell pole pieces, generally the negative pole. Due to the non-uniformity of the pole pieces, certain areas of the poles are at different potentials and act like small short-circuited chemical cells. This action sets up an internal current flow, which, though small, will eventually discharge a secondary cell or use up the service capacity of a primary cell. This local action is dependent on storage temperature. The higher the storage temperature, the more local action will occur. In some cells, this action is so strong that cell life may be reduced by 50% by storage at 30 °C (only 10 °C above room temperature). In contrast, storage at 10 °C (40 °F) may cut the loss during storage to half its room temperature value.

For this reason, it is a good idea to store carbon-zinc flashlight batteries under cold, but not freezing temperatures. This is best done by placing the cells in a sealed plastic bag for storage. Before the cells are used, they should be allowed to warm up gradually to normal room temperature, keeping them still sealed in the plastic bag to prevent water condensation inside the cells which could damage them.

C. Primary Cells

Nearly all the primary cells in use today are of the *dry cell* type like those developed during the nineteenth century by Georges Leclanche. Basically, the dry cell contains a water-solution electrolyte which is absorbed by a porous material, or mixed with a powder to form a moist paste. This is how we derive the term *dry* cell. Actually then, the term *dry* is relative; if the cell loses too much of its water, and truly becomes a dry cell, the electrochemical process will cease. It is inaccurate, in a technical sense, to call these cells *primary* cells. This is because roughly half the cells in this class can be recharged a limited number of times under carefully controlled conditions. However, this is not a good practice. The following paragraphs provide a detailed discussion of the construction and characteristics of some of the primary cells which the avionics technician may find in airborne, ground, or test equipment.

1. The carbon-zinc cell

The modern Lechanche, or *carbon-zinc cell* (or battery as it is usually called), differs only slightly in construction from the cell patented by Lechanche in 1860. Fig. 9-4 is a cutaway view showing the internal construction of a typical cell. The outer steel case is a relatively modern improvement. Formerly, the zinc can, which forms the negative pole, served as the outer casing for the cell. This was a particularly unsatisfactory arrangement that led to the destruction of numerous flash-lights and other devices, when sufficient zinc dissolved to release the corrosive, gummy, electrolyte paste into the battery compartment.

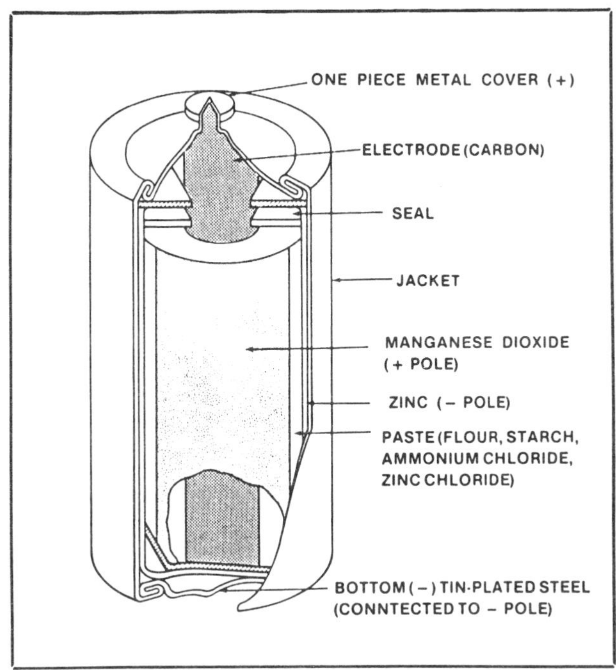

Fig. 9-4 The Leclanche (carbon-zinc) cell was invented by George Leclanche and patented in 1860.

As seen in Fig. 9-4, the electrolyte is composed of a water solution of ammonium chloride and zinc chloride mixed with starch and flour to form a gummy separator paste used to line the zinc can. The positive electrode of the cell is an electolyte-wetted mixture of manganese dioxide and powdered carbon packed around a carbon rod. Only the manganese dioxide powder of the positive electrode actively enters into the chemical reaction. The powdered carbon and carbon rod provide a low resistance connection to the powdered positive electrode and absorb some of the gas generated during cell operation.

NOTE: the manganese dioxide positive electrode also serves to absorb generated gas.

The actual chemical reactions that take place in a carbon-zinc cell are more complex than one would normally assume. But, for the avionics technician, it is enough to say that at the negative pole (the surface of the zinc can), metallic zinc combines with chlorine from the electrolyte in the cell to form zinc chloride, a white powder. At the positive pole, manganese dioxide is converted to manganous oxide. The result of the reaction is to transfer two electrons from the positive pole of the cell to the negative pole, using up the zinc, manganese dioxide, and ammonium chloride in the process.

Regardless of its size, the room-temperature potential difference across the poles of a new Leclanche cell with no load attached is 1.5 volts, but it should be noted that the potential difference across even a *new* cell depends upon the temperature. At 0 °F (−17.8 °C), the potential difference across a new cell will be 1.2 volts.

The service capacity of a chemical cell is directly related to its physical size. Since the Leclanche cells were the first to become popular, their standardized sizes have, to a great extent, determined the shapes and sizes of other types of cells. Table 9-1 gives the size and average service capacity, which may vary slightly from manufacturer to manufacturer, of the most commonly used Leclanche cells. As we shall see below, the disadvantages of carbon-zinc cells practically rule out their use in flight line equipment, but their continued use in test equipment such as multimeters justifies their inclusion here.

Cell Size	Length (in)	Diameter (in)	Average Service Capacity a-h at 20°C
AAA	1-11/16	25/64	0.2
AA	1-7/32	1-7/8	0.6
C	1-13/16	15/16	1.6
D	2-1/4	1-1/4	4.0
F	3-7/16	1-1/4	7.0

Table 9-1

One of the greatest disadvantages in the use of carbon-zinc cells is the shape of their discharge curve. Fig. 9-5, the graph of the cell potential difference against the time the cell has been in use, shows a steady decline in cell potential until the end-of-life point is reached.

This means that if a circuit is designed to operate with a potential difference of 6 volts produced by four Leclanche cells, the circuit will need periodic readjustment as the potential difference of the cells drops. In a few cases, this is only a minor inconvenience, as in the multimeter mentioned earlier. This is particularly true if the current in the circuit powered by the cell is small (under a few milliamperes), and the cell only furnishes power intermittently, with long rest periods be-

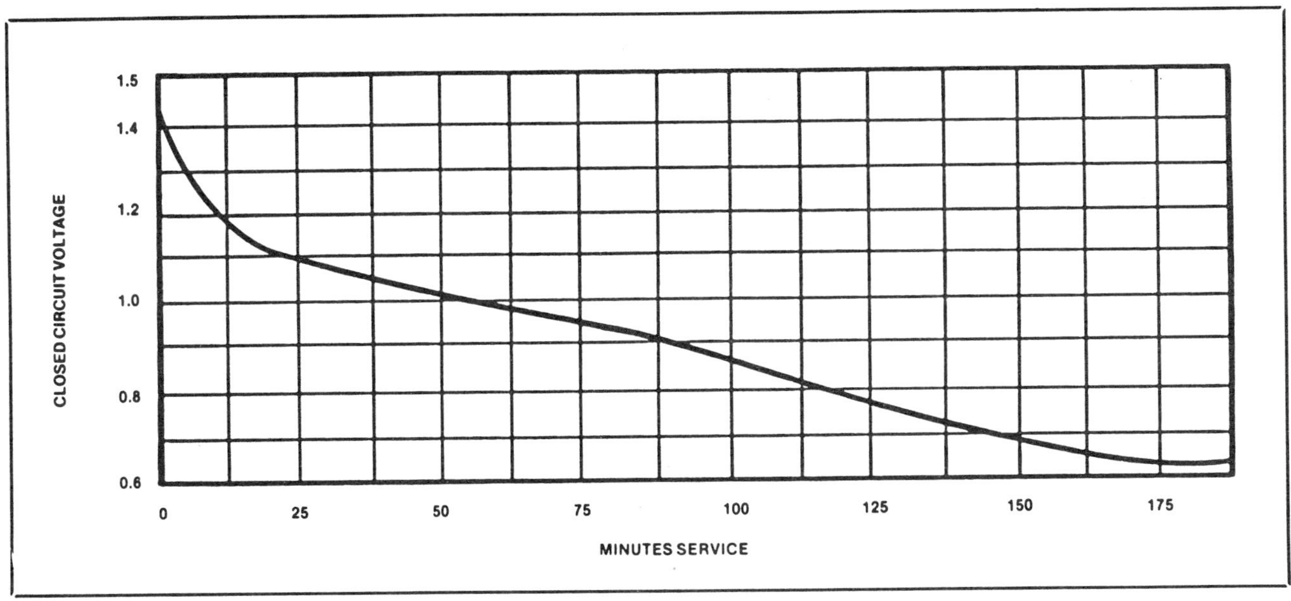

Fig. 9-5 Cell potential difference of size D carbon-zinc cell discharged at a continuous current starting at 667 milliamperes.

tween periods of use. Table 9-2 shows the significantly greater service capacity of a typical carbon-zinc cell when the circuit current is small. If greater circuit currents are required, it may be practical, as well as less expensive in the long run, to use a larger cell or several cells in parallel.

SERVICE CAPACITY VS CIRCUIT CURRENT (Intermittent Service) for a Typical Size D Cell	
Circuit Current (Amps)	Service Capacity to a 1-Volt Level
0.15	1.5
0.1	3.0
0.075	3.75
0.050	3.9

Table 9-2

The fact that Leclanche cells provide a greater service capacity when they are used intermittently, is due to the function and construction of these cells. When energy is drawn from a cell, gas is generated in the cell and the electrolyte in the neighborhood of the zinc can is used up. Tiny gas bubbles forming on the particles of the manganese dioxide positive pole serve to insulate the particles and slow or stop the chemical action that permits the electron transfer in the cell. This phenomenon is known as *polarization*. Similarly, a thin layer of depleted electrolyte next to the zinc can will slow or even stop ionic electron transfer. However, if the cell is allowed to rest after a short period of use, the gas will be absorbed by the powdered carbon and manganese dioxide, restoring the activity of the manganese dioxide particles. This phenomenon is known as *depolarization*. At the negative pole, diffusion of non-depleted electrolyte to the neighborhood of the zinc can will restore a fair portion of the former zinc pole area to use. Thus, the voltage across the cell will begin to drop as energy is drawn from the cell and polarization and depleted electrolyte layering occur. But the power output will return to a higher level, if the cell is allowed to depolarize during a rest period.

The ability of the zinc-carbon cell to recuperate during rest makes this type of unit useful for certain kinds of intermittent operation. In fact, the service life of a Leclanche cell may be doubled if it is used intermittently. Other types of cells show a much smaller improvement in service life because of intermittent operation. For example, alkaline cells show only a 20% lengthening of service life when used intermittently, and mercuric oxide cells show none at all. For mercuric oxide cells, the life span for intermittent and continuous low-current service are almost identical. This does not mean that the service capacity of the Leclanche cell is increased by intermittent operation. It means this sort of operation will permit use of almost *all* the service capacity, and that the service life of the cell may be made to approach the shelf-life of the unit.

Another drawback to the use of Leclanche cells in avionics is their sensitivity to low temperatures. Although they are not as sensitive as mercuric oxide cells, which become practically useless at temperatures of 20 °F (−6.7 °C), a carbon-zinc cell should not be connected to a circuit at temperatures below 20 °F (−6.7 °C) or above 130 °F (54.5 °C). Operation beyond these extremes will cause a quick or even immediate deterioration of the cell.

A further drawback to the use of Leclanche cells is their limited shelf life. After two years of storage at room temperature, the cell is generally no longer serviceable, even if it was never used. For many applications, however, the low cost and compactness of Leclanche cells outweighs the disadvantages.

2. *The zinc chloride cell*

The *zinc chloride cell* is an improved version of the carbon-zinc cell. The differences between the two arise from the fact that zinc chloride is substituted for the Leclanche mixture of zinc chloride and ammonium chloride. This seemingly minor change results in significant differences in some of the operating characteristics of the cells. Fig. 9-6 is a cutaway view of a typical zinc chloride cell showing the considerably more complex seal required by this type of cell. This improved seal is required because of the different role that water plays in the chemistry of the zinc chloride cell.

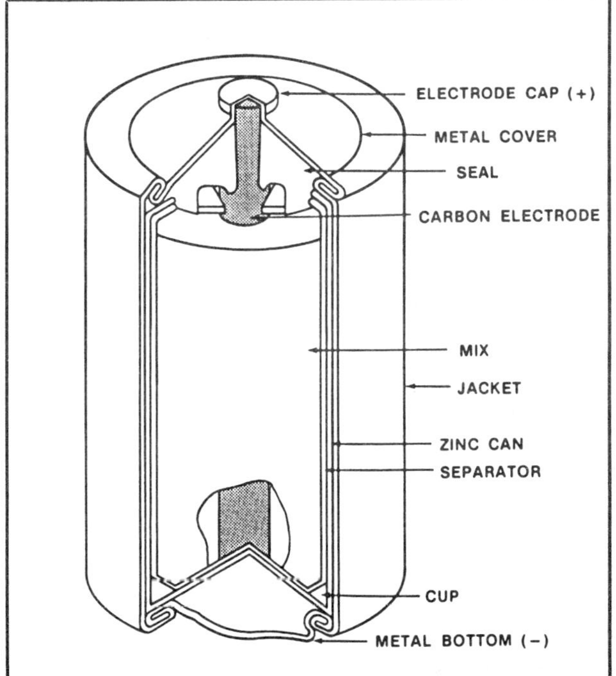

Fig. 9-6 Note the close resemblance to the Leclanche cell in this view of a typical zinc chloride cell.

At the positive pole, manganese dioxide is converted to a hydroxide by the addition of hydrogen which is supplied by the water in the cell. At the negative pole, metallic zinc is converted into zinc oxide and zinc chloride, and a small amount of water is formed. It is important to point out that more water is broken up at the positive pole than is produced at the negative pole, so that the electrolyte solution gradually dries out as the cell is used. The zinc chloride cell is quite dry when the end of its service life is reached.

The lack of a starch paste, the combination of hydrogen and oxygen at the poles of the cell, and the increased diffusion rate of the zinc chloride electrolyte in the zinc chloride cell make it possible to obtain a greater current flow rate from a cell of this type without experiencing the gas formation and layer of depleted electrolyte found in the standard Leclanche cell. In general, practical current flow rates as much as 50% higher may be expected from zinc chloride cells without a shortening of service life. Although the discharge curve of the zinc chloride cell is sloped like that of the Leclanche cell, the service capacity of zinc chloride cells is about 50% higher than for a carbon-zinc cell of the same size.

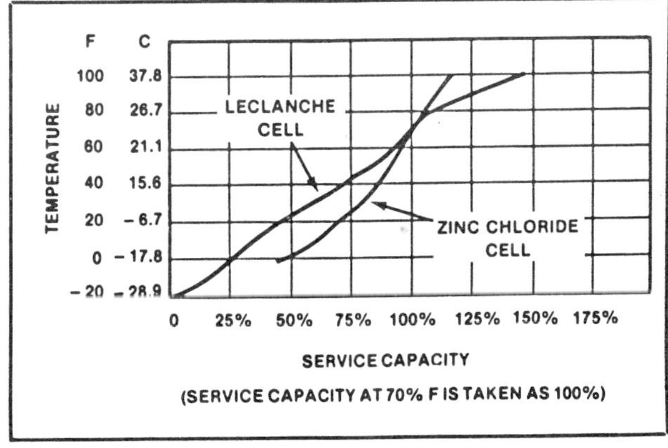

Fig. 9-7 Relative service capacities of Leclanche and zinc chloride cells at differing temperatures.

The zinc chloride cell also provides better low temperature service than the Leclanche cell (Fig. 9-7). Although it is somewhat more expensive than the Leclanche cell, the zinc chloride cell presents fewer drawbacks than the older type. According to manufacturers' specifications, zinc chloride cells may be stored at temperatures ranging from −40° to 160°F (−40° to 71.1°C) and may be used at temperatures ranging from 0° to 160°F (−17.8° to 71.1°C).

3. The alkaline cell

Another recent improvement of the Leclanche cell is the so-called *alkaline cell* (alkaline-manganese dioxide would be a better name, since many cells use an alkaline solution as the electrolyte.) The electrolyte used in the alkaline cell is a water solution of potassium hydroxide, commonly known as lye. This, coupled with the construction differences, results in a high current cell with good low temperature performance. Fig. 9-8 illustrates the physical design differences between the alkaline and Leclanche cells. First of all, the solid zinc can which formed the negative pole has been eliminated. A layer of powdered zinc is used, since the zinc particles provide a greater surface area for the electrochemical reaction than a piece of solid zinc. Contact with the powdered negative pole is supplied by a hollow brass cylinder which contains additional electrolyte to counteract loss through evaporation or through gas formation while the cell is in service. As in the Leclanche cell, the positive pole is formed of a powder mixture using manganese dioxide as the active ingredient.

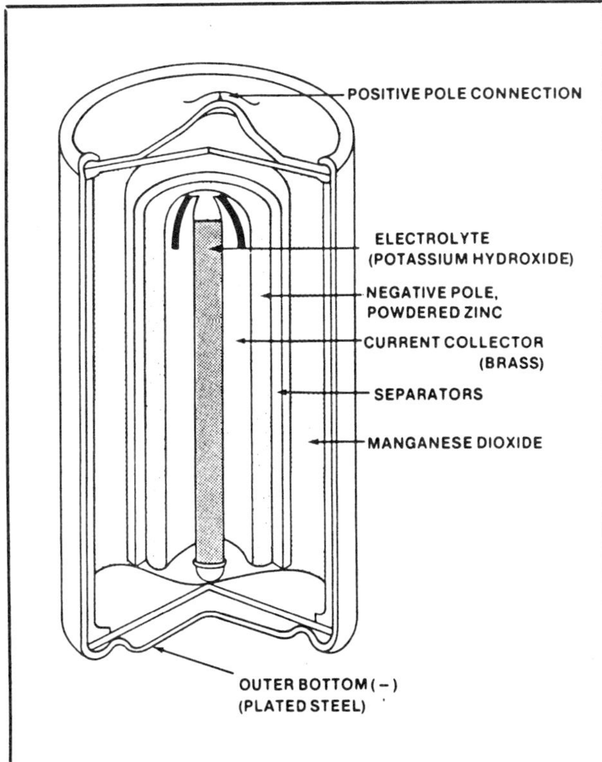

Fig. 9-8 Structurally quite different from the Leclanche cell, the alkaline cell uses high surface area powdered materials for both poles.

Although the chemical process of the alkaline cell is different from that of the ammonium chloride and zinc chloride electrolyte cells, the potential difference between the poles of a new alkaline cell is still 1.5 volts, since the pole materials involved in this cell are still manganese dioxide and zinc. At the positive pole, manganese dioxide gives up some oxygen and is changed into another oxide of manganese. At the negative pole zinc is converted into zinc oxide. Neither the potassium hydroxide nor the water in the cell are affected by the reaction though both enter into it.

Due to its construction, the alkaline cell has a much greater ability to provide high current service, without losing its initial service capacity, than either the Leclanche or zinc-chloride cells. This is in part because the effects of cell polarization are minimized by the large effective surface area of the powdered zinc negative pole. This type of cell offers both an increase in service capacity over other manganese dioxide-zinc cells, and the possibility of utilizing most of this service capacity in continuous high-current operation.

A comparison of Leclanche, zinc chloride and alkaline cells, shows the superiority of the alkaline cell. Under conditions of relatively high current flow, an alkaline cell may provide up to seven times the service life of the familiar Leclanche cell. Although the service capacity of the alkaline cell is only double that of the carbon-zinc cell, the alkaline cell is much more likely to deliver a higher proportion of its capacity than the Leclanche cell.

Both the low-temperature performance and the shelf life of the alkaline cell are considerably better than the Leclanche cell and somewhat better than the zinc chloride cell. The high temperature characteristics of alkaline cells are about the same as Leclanche cells, and somewhat poorer than zinc-chloride cells. Operating temperatures according to manufacturers' specifications range from −20° to 130°F (−28.9° to 54.4°C). Storage is possible within a range from −40° to 120°F (−40° to 48.9°C). Alkaline cell shelf life at normal room temperatures is estimated to be about four years or twice that of Leclanche cells under similar conditions.

4. Mercuric oxide and silver oxide cells

Until quite recently, the *mercuric oxide* or *mercury* cell, was the most stable and most compact primary cell available. Although expensive, it was a natural choice for miniature equipment, since mercury cells offer nearly three times the energy per unit weight as Leclanche cells, and four times the energy per unit volume. Chemically, the mercury cell is like the alkaline cell, except that mercuric oxide is substituted for the manganese dioxide positive pole used in the alkaline cell. When the cell is in use, the mercuric oxide positive pole is *reduced*, that is, oxygen is removed from every molecule of mercuric oxide. At the positive pole, zinc is *oxidized* to form zinc oxide. Although the water solution of potassium hydroxide (lye) which serves as the electrolyte enters into this process, it is not used up. The mercuric oxide and zinc, on the other hand, are converted by the mercuric oxide cell; in fact, 80 to 90% of the material is converted. This is the reason for the long service capacity of the mercury cell.

Fig. 9-9 shows the physical structure of the more popular kinds of mercury cell. There are two kinds of "button" cell available. In the "button" cell of Fig.9-9A, the potassium hydroxide electrolyte is held in place by absorbent pads. The gelled negative pole cell shown in Fig.9-9B immobilizes the potassium hydroxide electrolyte by mixing it with the powdered zinc and a gelling compound. Button cells are most often used in hearing aids, photographic equipment, or as voltage references in measuring equipment where the circuit current requirements are small.

The cylindrical mercury cell pictured in Fig.9-9C is manufactured in the same sizes AA, C, and D as are Leclanche, zinc chloride, and alkaline cells. The most important difference is that the top cap *is the negative pole* in most cylindrical and button type mercury cells. A further difference between manganese dioxide-zinc cells and mercury cells is the potential difference across the cell. All new manganese dioxide-zinc cells show a potential difference of 1.5 volts. A new mercury cell with a positive pole composed only of mercuric oxide and carbon will produce a potential difference of 1.35 volts. A number of manufacturers also produce a mercury cell with a positive pole composed of a mixture of mercuric oxide, manganese dioxide, and carbon, which produces a potential difference of 1.4 volts. The 1.35 volt cell

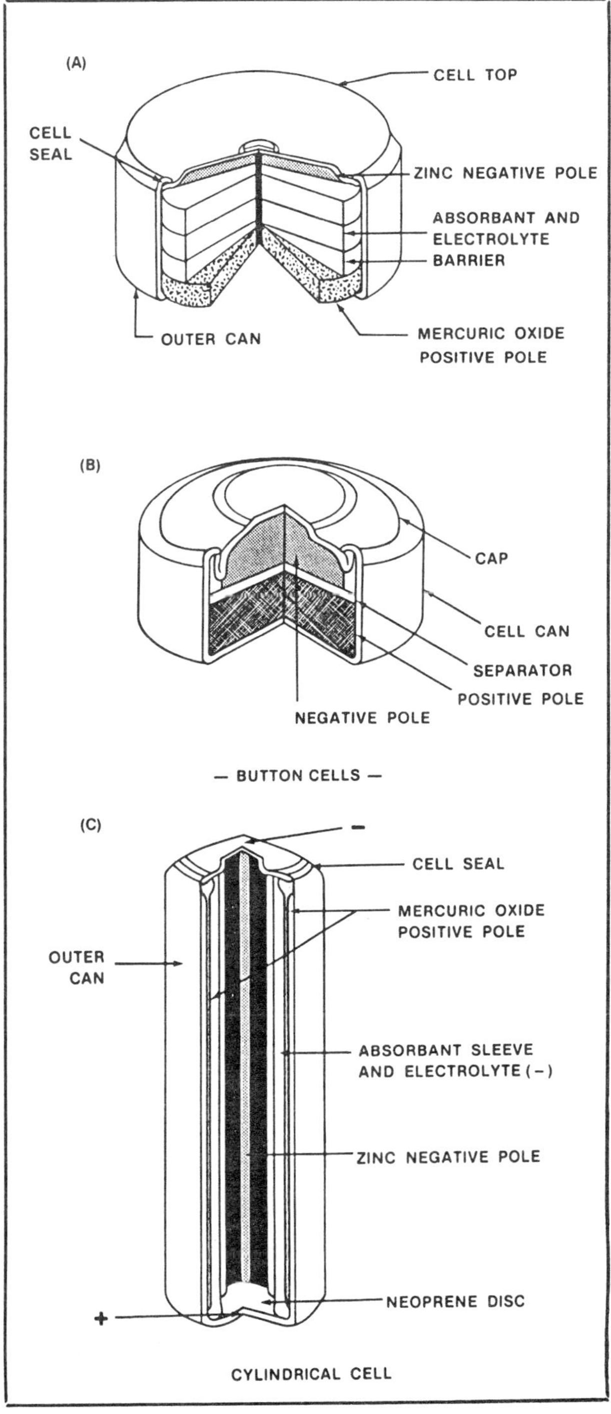

Fig. 9-9 Popular mercury cell types.

shown in Fig. 9-9 is perhaps the more interesting for avionics technicians, because it has a flat discharge curve. This means that the potential difference across the cell remains fairly constant until the cell is used up. At that time, the cell voltage drops sharply. This characteristic makes the mercury cell useful for powering portable test equipment that requires a constant input voltage or a voltage reference.

Although it is not capable of functioning at high circuit currents like the alkaline cell, the mercury cell can provide a service capacity nearly 50% greater than the alkaline cell, 2.4 a-h for a size AA cell and 14 a-h for a size D cell. The shelf life of the mercury cell is also better than that of the manganese dioxide cell. Two to four years is the average.

Closely related to the mercury cell is the *silver oxide cell.* This cell is much the same as the mercury cell, except it uses silver oxide in place of mercuric oxide as its positive pole. Its open circuit output voltage is 1.60 volts, slightly higher than that of the Leclanche system. Neither the silver oxide nor the mercury cell is in wide use in avionics or applied applications, because neither cell is capable of functioning well at low temperatures. The silver oxide cell is reported to be better in some manufacturers' specifications, but neither is recommended for use below 32 °F (0 °C).

D. New Types of Primary Cells: Magnesium and Lithium Cells

A few years ago, experiments were carried out with Leclanche-type cells using magnesium metal as the negative pole. The problems encountered made it necessary to use special electrolytes composed of bromine or magnesium and chlorine. Eventually, a practical *magnesium cell* with good high current capacity, service capacity, and excellent shelf life was developed. For a while it looked as if this new cell type was going to replace the manganese dioxide and mercury types, but the magnesium cell now seems to have been pushed into the background by the newer lithium cells. Due to its very long shelf life, light weight, and very high energy content, the lithium cell is beginning to find a place as a back-up power source for avionics computer circuits. This is particularly the case for computer memory circuits, where even a momentary loss of power may result in the erasing of much valuable information.

Rather than speak of one type of lithium cell, it is better to speak of a family of cells, all of which use the element lithium (the lightest metallic element), as their negative pole. There are presently more than a dozen different lithium cell chemical systems in production. Each of them requires a different physical structure. For this reason, no "typical" cell construction details will be discussed here. Only those characteristics common to all, or most of the currently produced cells will be included.

All lithium cells produce a potential difference which is considerably higher than that of any other commercial primary cell chemical system. The two most popular systems utilize a lithium negative pole and a positive pole composed of a gas dissolved in the electrolyte. They produce open-circuit potential differences of 2.8 and 3.6 volts respectively. This is twice that obtained from any manganese dioxide cell. The discharge curve (after a short break-in period during which the voltage across the cell *increases*), is quite flat. The lithium cell will produce its rated potential difference almost until it stops functioning completely.

Besides being higher than that of any other cell, the service capacity of lithium cells seems unaffected by use in circuits requiring a high current flow. A D-size cell which is rated at 8.0 a-h will deliver 0.8 A for 10 hours, 80 mA for 100 hours, or 80 mA for 10^5 hours, until it ends its useful life with a potential difference of 2 volts across its poles. Table 9-3 gives the service capacity of the more common size cells. Remember, since the potential difference across a lithium cell remains roughly twice that developed by a Leclanche-type cell, the power (P = E × I) delivered by a lithium cell will be twice that delivered by a Leclanche cell at the same current flow.

Cell Size	Capacity in Ampere-Hours	Discharge Current to 2 Volts 10 Hours	100 Hours	1000 Hours
AA	1.2	120mA	12mA	1.2mA
C	4.0	400mA	40mA	4.0mA
D	8.0	800mA	80mA	8.0mA

Table 9-3

Lithium cell performance at extreme temperatures is one of the characteristics that make them desirable for use in airborne equipment. The cell will function at temperatures as high as 165 °F (73.8 °C) with full voltage output and full service capacity. At low temperatures, its performance is greatly superior to even the magnesium cell. At −20 °F (−28 °C), a temperature at which Leclanche, alkaline, and mercury cells will produce no output, the lithium cell will deliver up to 96% of its rated service capacity. Even at temperatures of −40 °F (−40 °C), the lithium cell will deliver 60% of its capacity.

Equally important for its use as a standby power source in navigation or communication equipment or as the primary source in emergency locator beacons, the lithium cell has an extremely long shelf life and is much more tolerant of storage under high temperature conditions than any other type of primary cell. Because there is very little local action, a lithium cell may be stored for up to ten years at a temperature of 68 °F (20 °C) with a loss of less than 25% of its rated service capacity, or at a temperature of 130 °F (54.4 °C) for eight years with a similar loss of only 25% of its service capacity.

Besides its cost, which is still relatively high, the principal problem in the use of lithium cells in avionics is the care that must be used with a material as potentially dangerous as lithium. Like phosphorus, it will burn when exposed to air.

E. *Secondary Cells*

Although the lithium cell is in the process of obtaining more acceptance as a standby power source, the secondary, or rechargeable cell is still the most widely accepted type of unit. Unlike the primary cells which are used to *generate* electrical energy, these cells are used to *store* electrical energy. The chemical process which goes on inside the cell is reversible, that is, by applying a potential difference, derived from some other source, across the poles of the cell and forcing a *charging current* through it, we may return the poles and electrolyte to their original or near-original condition. At present, every aircraft, except gliders, and almost every ground installation is equipped with either a lead-acid battery formed of Plante cells or a nickel-cadmium battery. Even those ground installations which have motor-driven auxiliary generators require the presence of secondary cells to start the generator drive motor.

The majority of the secondary cells which the avionics technician will presently see in service are the large, vented, *wet* cell types. These cells are called *wet cells* because the electrolyte is a liquid. This liquid is either sulphuric acid and water in the case of the Plante cell, or potassium hydroxide and water in the case of the ni-cad cell. The cell has a vented plug which is designed to permit the escape of gas formed during the charging or discharging processes and to prevent the spilling or escape of the electrolyte solution. These plugs may also be removed for adding additional electrolyte or for testing the concentration of the electrolyte solution. Since the theory of operation, construction, and maintenance of batteries formed of these cells has been thoroughly covered in an Aviation Maintenance Publishers training manual, *Aircraft Batteries (Lead-Acid — Nickel-Cadmium)*, the reader may wish to refer to that manual for information concerning these larger, unsealed cells. For the most part, our discussion will concentrate upon the smaller, sealed cells which have been developed in recent years for use in portable equipment or as back-up power sources.

1. *The Plante lead-acid cell*

The lead-acid, or *Plante storage cell* has proved itself so successful as a power source that only within the last couple of years has there been any significant improvement over the early cell and battery designs. All lead-acid cells use the same chemical system. A positive plate of lead dioxide and a negative plate of porous sponge lead are suspended in a solution of water and sulfuric acid. At the positive pole, the lead dioxide reacts with the sulfuric acid to produce lead sulfate. This releases negative oxygen ions to the solution. At the negative plate, sponge lead is converted into lead sulfate. The oxygen from the positive pole unites with hydrogen in the electrolyte to form water. This action dilutes the electrolyte solution. The process produces a potential difference of about 2 volts across the poles of the cell and will provide energy to an external circuit at a gradually decreasing voltage until the positive and negative poles are coated with lead sulfate and the electrolyte solution becomes too dilute to supply sufficient chemical action. This is called the *discharged state.*

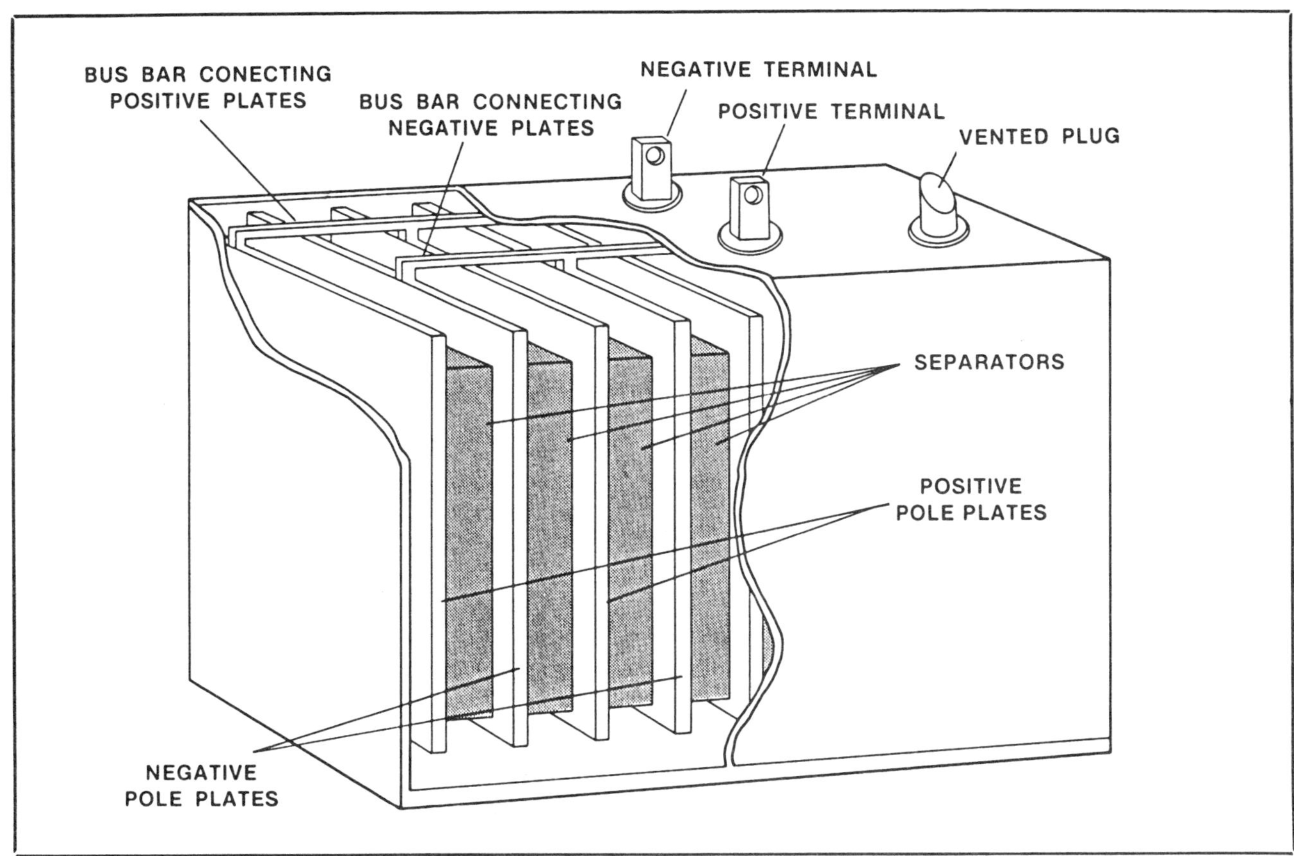

Fig. 9-10 Cutaway view of a Plante lead-acid cell.

When a Plante cell has reached its fully discharged state, or before, when it is only partially discharged, a *charging current* may be passed through it in the direction opposite to the normal current flow through the cell. This charging current has the effect of reversing the chemical activity of the cell. At the positive pole of the cell, lead sulfate is converted to lead dioxide by taking oxygen from the water in the electrolyte and returning sulfate ions to it. At the negative pole, sulfate ions are also returned to the electrolyte solution. The charging process, if carried on long enough, will eventually remove all the lead sulfate from the pole plates, restoring the cell to its original state.

If carried out with a high charging current, the process is accompanied by the generation of heat and gas at both poles. When there is no more lead sulfate to convert, the charging current begins to electrolyze the water in the cell electrolyte. This means that the water is broken down into hydrogen and oxygen. Continuous charging such as is done in many cases when the cells are in standby service, may seriously deplete the amount of water in the cell, cause serious overheating, or create a potentially explosive mixture of hydrogen and oxygen in the battery enclosure.

As shown in Fig. 9-10, a Plante cell is composed of a number of interleaved positive and negative plates, with separator sheets between them, suspended in the electrolyte solution. By connecting all the positive plates together and all the negative plates together, the equivalent of a cell with a large plate area is produced. Since the service capacity of the cell depends upon its plate area, close packing of a large number of plates will produce a cell with a large energy to volume ratio. One manufacturer produces cells ranging in size from a small unit measuring 3 × 5 × 10 inches (7.6 ×12 × 25.4 cm) with an a-h capacity of 17 a-h when discharged over 8 hours to an end voltage of 1.75 volts.

The potential difference across a Plante cell decreases gradually as the service capacity is used. This presents less of a problem in the Plante cell than in the case of the Leclanche cell, since the cell may be recharged before the voltage falls to a level that would require readjustment of the circuit to which the cell is connected. Although many motorists may disagree, the lead-acid cell exhibits good low temperature performance when the circuit current requirements are moderate. The difficulties sometimes experienced with lead-acid motor-starting batteries arises from the fact that temperature effects are much more pronounced at high current flows than at moderate flows. Loss of charge in a cell not connected to an external circuit is also more pronounced under high or low temperature conditions, but the Plante cell stands up well to low temperature storage. High temperature storage conditions, on the other hand, may have a destructive effect upon a lead-acid cell, and storage temperatures above 130 °F (54.4 °C) should be avoided.

2. The sealed lead-acid cell

Recently, automobile battery manufacturers have started to market units which they advertise as "sealed, maintenance free" batteries. These units have no provision for the addition of water to replace the water lost by normal evaporation or through overcharging. Instead, a sufficient amount of extra water is designed into the electrolyte system so that the addition of water is not required through the projected life of the battery. Although these units represent an increase in convenience over conventional Plante batteries, they should not be confused with *true* sealed lead-acid cells.

The modern sealed lead-acid cell is usually marketed in cylindrical form, in the *D* or somewhat larger *X* size. The internal construction of such a cell is pictured in Fig. 9-11. Thin lead dioxide and sintered lead strips sandwiched with porous separator strips are rolled to produce a cell with a large effective plate area in a relatively small volume. The sulfuric acid and water electrolyte is absorbed by the separator material and is thus immobilized, so that the cell may be used in any position.

Since the cell is completely sealed (except for a safety vent), the gas formed during the charging process cannot escape. Instead, it is recombined to form water. The cell acts as a closed system, and no water is lost. Of course, the charging rate must be carefully controlled so that the rate of

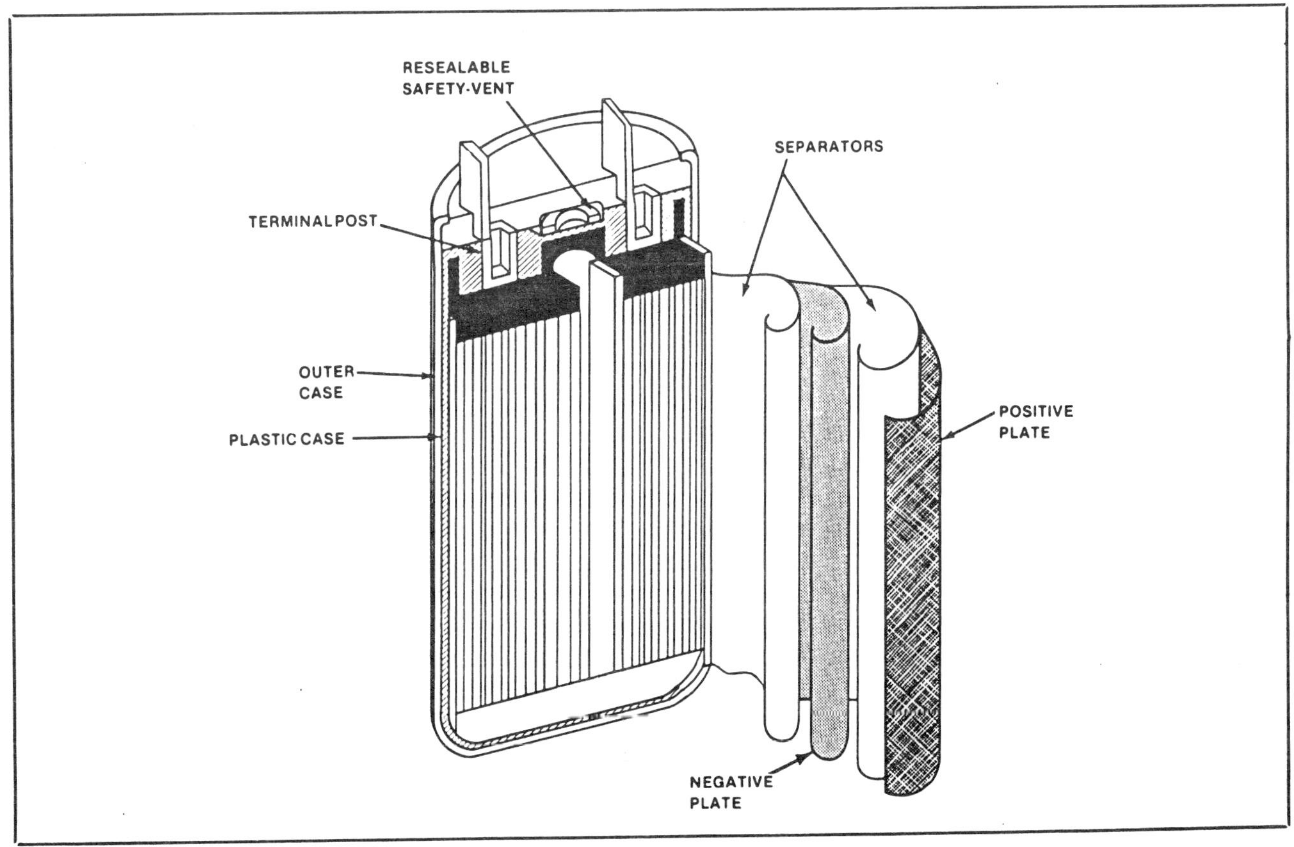

Fig. 9-11 A cutaway view of a sealed lead-acid cell.

gas formation does not exceed the cell's ability to recombine the gas into water. If the charging rate is too high, some gas will be vented through the safety vent, preventing the cell from bursting. Table 9-4 lists the dimensions and service capacities at 20 °C of the two most popular sealed lead-acid cell sizes.

Cell Size	Length	Diameter	Service Capacity a-h (20 hrs)
D	2.4 in. (6 cm)	1.34 in. (3.4 cm)	2.7
X	2.85 in. (7.2 cm)	1.74 in. (4.4 cm)	5.2

Table 9-4

Like the Plante cells whose chemistry they utilize, sealed lead-acid cells are capable of providing very high currents (on the order of 100 amperes from a D sized cell), for very short periods of time, or small currents for a long time, without permanent damage to the cell. After the service capacity is used up, the cell should be recharged immediately. Cell damage in any lead-acid cell can occur if the discharged cell is left connected to an external load without recharging.

The low temperature operating characteristics of sealed lead-acid cells are good, and they may be expected to produce some useful output at temperatures as low as −40 °F(−40 °C). The lowest practical temperature for Plante cells is about −20 °F. The service capacity of the cell increases with the temperature and the cells may be used in surroundings with temperatures as high as 140 °F (60 °C).

Storage temperatures are about the same as the operational temperatures, with low temperature retarding the self-discharge or loss of charge while "on the shelf" At room temperature, a sealed lead-acid cell may be expected to retain a safe level of charge for about three years. The cell should always be recharged before this time has elapsed, in order to maintain at least a safe level. This will prevent lead sulfate crystals, which reduce the service capacity or destroy the cell, from forming when the charge level falls below the safe minimum point.

Unlike the case of many primary cells, where the decrease in the potential difference across the cell is a disadvantage, the sloping discharge curve of the sealed lead-acid cell is a very useful characteristic. In the Plante cell, it is a normal practice to measure the density, or rather the concentration of sulfate ions in the electrolyte solution. This tells us how much of the sulfate has been used up to form lead sulfate on the cell plates and gives us a good idea of how much service capacity is left in the cell. This measurement tells us if it is necessary to recharge the cell.

There is no way to get at the electrolyte of a sealed lead-acid cell. It is inside a sealed unit and absorbed in the porous separator. Only the potential difference across the cell may be used to determine the state of the charge. Fig. 9-12 illustrates how the potential difference across a sealed lead-acid cell can indicate the percent of service capacity left in the cell. Although the accuracy of this graph is fairly poor, it may be used to roughly determine when sealed lead-acid cells need to be recharged.

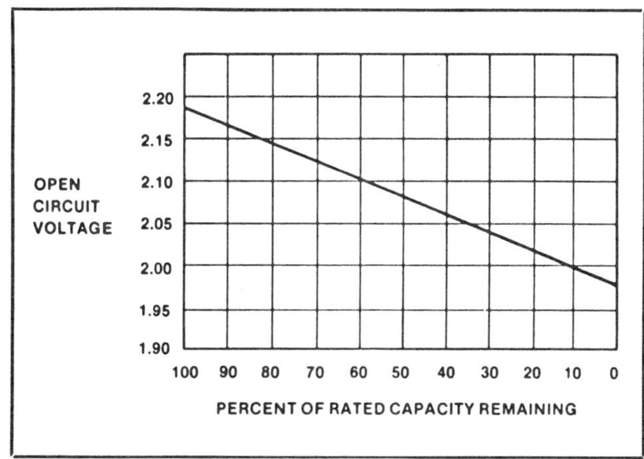

Fig. 9-12 Percent of rated capacity remaining in a sealed lead-acid cell as shown by the cell potential difference.

There are various ways of recharging a sealed lead-acid or Plante cell. In general, any method that may be used for the Plante cell is also acceptable for use with the sealed unit. If the cells are used as the only power source, as in portable equipment for example, the fastest safe method of recharging the cells might be best. This would be to connect a constant voltage across the cell in such a way as to force a charging current through it. The charging current must flow in the direction *opposite* to the direction in which the current normally flows through the cell when it is in use. This may be accomplished by connecting the positive potential to the positive pole of the cell, and the negative side of the charging potential to the negative pole. A charging potential of 2.55 to 2.60

volts will charge a sealed lead-acid cell to 95% of its rated service capacity in a half hour. This is called the *fast charge* method. However, it is a brutal way to treat cells.

A charging potential difference of less than 2.4 but greater than 2.3 volts is called the *float charging potential.* At this charging potential, it takes about 16 hours to fully recharge a cell to 100% of its service capacity. This method puts much less strain on the cell, increasing its useful life by a factor of seven times or more.

The float charge method may be used in many units or installations where power is derived from commercial service lines. In units such as computer memories or beacons, where uninterruptable power is required, lead-acid or sealed lead-acid cells may be left across the power supply line with a float charge potential impressed across them. If the commercial service should temporarily fail, the cells would automatically begin to discharge through the external circuit, maintaining the power to essential circuits.

Another method of charging lead-acid or sealed lead-acid cells is the *constant current method.* This consists of forcing a constant current through the cell in the direction opposite to the direction of normal current flow. In most cases a current rate equal to the current which may be drawn for ten hours from a fully charged cell is used. This is called the *C/10 charge rate.* A totally discharged cell must be charged by a C/10 current for 14 to 16 hours in order to be returned to 100% of its initial service capacity. Constant current charging is particularly useful when a series-connected battery or a number of cells must be charged, since each cell in the series string will be charged to its capacity, regardless of the initial state of each cell.

Passing a small charging current equal to the 500-hour discharge rate or less through a cell is called *trickle charging.* This is the constant current equivalent of float voltage charging, and like float charging, it is used to maintain the charge on cells in standby service.

3. Nickel-cadmium secondary cells

Available in cell sizes from the tiny button and AA-sized cells to giant units equivalent to the lead-acid cells used on submarines, the nickel-cadmium, or *ni-cad,* cell represents the most versatile storage cell presently available. Larger, prism-shaped, vented cells like those pictured in Fig. 9-13 can be used singly or in grouped batteries in place of Plante cell lead-acid batteries. In this type of cell, the positive and negative poles are in plate form, as in the Plante cell. The pole plates may be made of an active material metallic cadmium in the negative plates and nickelic hydroxide in the positive, embedded in a very porous nickel matrix which serves to support the active materials and to provide electrical contact. The separators shown in Fig. 9-13 are generally made of a porous, non-woven fabric impregnated with a water solution of potassium hydroxide. This is the electrolyte in all ni-cad cells.

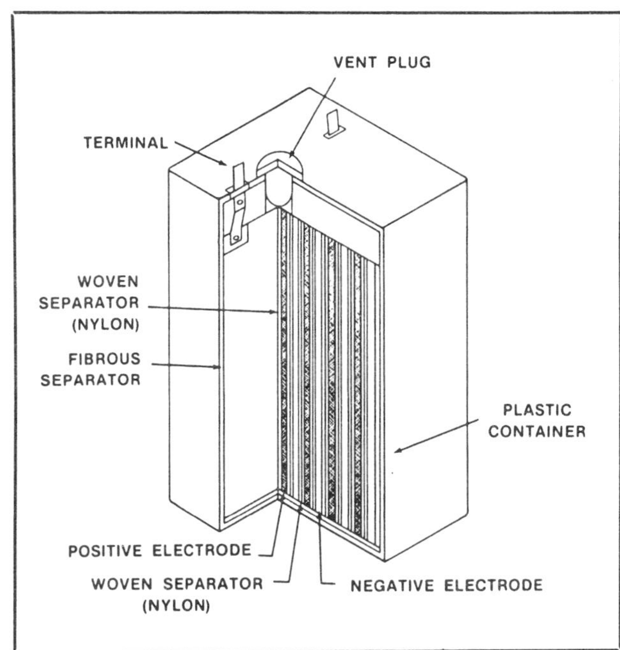

Fig. 9-13 A vented nickle-cadmium cell — a large heavy duty unit presently used in place of Plante cells in many applications.

Of greater interest to the avionics technician are the smaller button and cylindrical ni-cad cells which are used to power many portable test instruments and serve as standby sources in a number of equipment types. The construction details of typical button and cylindrical cells are shown in Figs. 9-14 and 9-15. As seen in the illustrations, these are sealed units equipped with resealing vents which permit the escape of gas formed by a too high charging rate. In the button cell, the pole pieces are formed of powdered cadmium and nickelic hydroxide retained by nickel gauze mesh or formed into small plates.

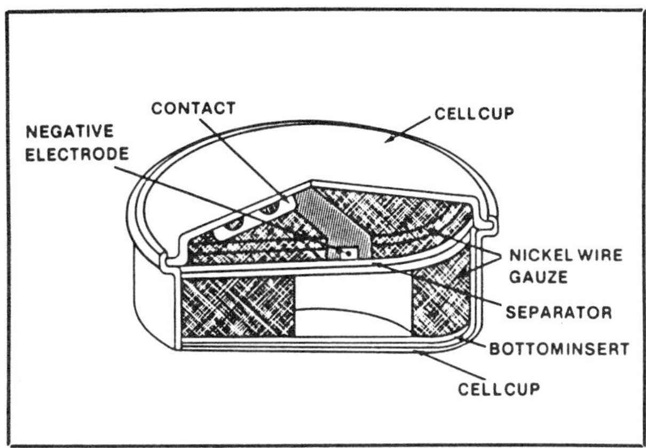

Fig. 9-14 Cutaway view of a ni-cad button cell.

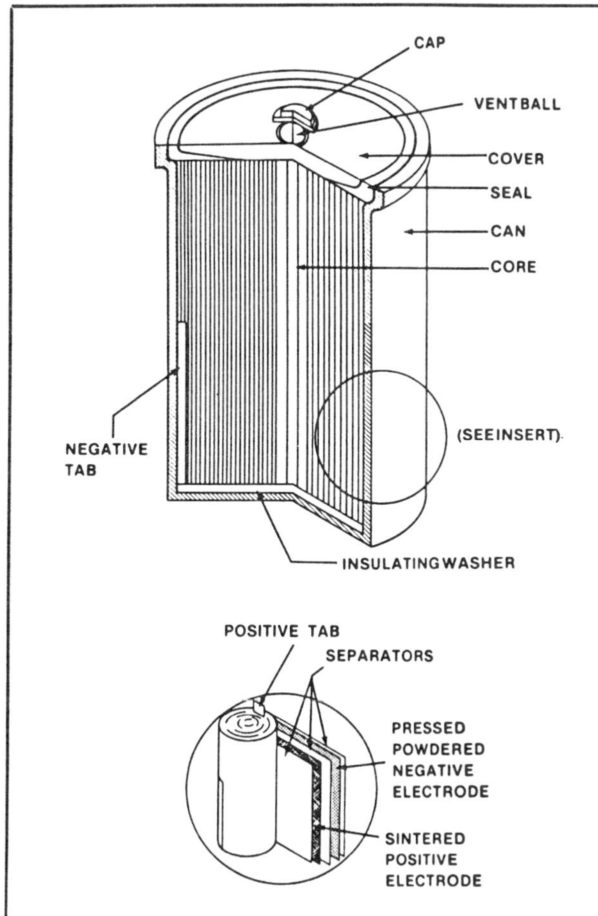

Fig. 9-15 View of a cylindrical type ni-cad cell.

In the cylindrical cell, the positive pole is formed by nickelic hydroxide embedded in a porous nickel ribbon. The negative pole piece is made of pressed cadmium powder. As in the case of the sealed lead-acid cell, the positive and negative poles are interleaved between separator strips and the whole is then rolled into a cylindrical shape. Regardless of its physical shape, the chemical system of the ni-cad cell involves the formation of cadmium hydroxide at its negative pole and the reduction of nickelic hydroxide to nickel hydroxide at its positive pole. This requires a certain amount of water to be broken down and fixed in compound during the discharging process. But this water is returned to the electrolyte when the cell is recharged.

The potential difference across the poles of a fully charged ni-cad cell is about 1.2 volts, or slightly more than half that obtained from a lead-acid cell. The service capacities and dimensions of some of the more popular sizes are given in Table 9-5. Of special interest is a small, encapsulated cell recently put on the market which is designed to be soldered directly to a printed circuit board to provide standby power to the circuits surrounding it.

Sizes of Common Nickel-Cadmium Cells

Size	Service Capacity a-h	Dimensions Diameter	Height
AA	0.450	0.56 in. (1.42 cm)	1.95 in. (5.04 cm)
C	1.200	0.88 in. (2.23 cm)	1.67 in. (4.24 cm)
D	4.000	1.30 in. (3.3 cm)	2.43 in. (6.17 cm)

Table 9-5

The ni-cad cell provides fairly good performance at both high and low temperatures, possessing an operating range, depending on type, from −4° to 113°F (−20° to 45°C), although the low temperature during charging should be kept above 32° (0°C) to ensure a full charge. The storage temperature range of ni-cad cells is quite broad, going from −40° to 140°F (−40° to 60°C). It should be noted, however, that high temperatures reduce the service capacity of ni-cad cells sharply both when in service and when in storage, due to loss of charge through local action. The self-discharge rate of ni-cad cells is higher than that of sealed lead-acid units. But the ni-cad cell will tolerate storage in a discharged condition much longer than lead-acid units.

Because of its light weight, reliable service, and good temperature characteristics, the ni-cad cell will probably be increasingly used in airborne equipment. Because of this, a number of potential problems associated with ni-cad cells should be recognized by the avionics technician. First among these is the problem of *thermal runaway*. This condition can occur when the cell is subjected to excessive charging rates. The build-up of heat in a charging cell which is nearing its 100% charge level will lower the cell resistance, thereby permitting a greater charging current through the cell. If this proceeds, the cell can overheat to the point of rupture or even explosion. Normally, however, the constant current or float voltage charging methods permit recharging without an excessive build-up of heat.

Another critical problem noted with ni-cad cells is the so-called *memory effect*. This condition occurs when a cell, particularly a cell used in standby service, is discharged a number of times to only a small percentage of its service capacity. For example, if a cell capable of delivering 5 a-h is required to supply only 2 a-h during a number of transient power line failures and is recharged to full service capacity after each use, the cell somehow "remembers" its earlier exertions, and will refuse to give more than the 2 a-h to which it has become accustomed. The only solution to this problem is a partial one. A technician has to completely discharge the ni-cad cell periodically. If too many shallow discharge cycles have not taken place between the deep discharge cycles, the memory effect will be minimized.

The third problem involves the use of ni-cad cells connected in series. This is a very normal situation because of the low potential difference (1.2 volts) across a single cell. As a battery composed of a number of cells is used, the cell with the lowest remaining service capacity will discharge first. If the battery continues in service, this cell will begin to charge up with a reverse polarity, generating large quantities of potentially explosive hydrogen gas. This condition may be avoided by periodically discharging *all* the cells in the battery, then charging them all up to 100% service capacity. Also, when doing this, make sure that all the cells in a battery are of the same age and in roughly the same physical condition.

The fourth and last consideration in the use of ni-cad cells is both an advantage and a disadvantage. This is the rather flat voltage discharge curve shown by these cells. A ni-cad cell will show much the same potential difference across its terminals, until the cell service capacity is totally depleted. Although this makes it unnecessary to readjust circuits or to provide external regulation of the cell's output, it also makes it impossible to determine just how much service capacity is left in the cell. For this reason, ni-cad cells are best put in a float or trickle charge state that will bring them up to full charge after each use.

The ni-cad cell is more sensitive to damage through faulty charging than the Plante, or sealed lead-acid cell. In all cases, it is best to refer to manufacturers' recommendations for that particular type of cell before attempting to recharge it.

A. Self-test Questions:

State whether the following sentences are true or false. If the sentence is false, rewrite the sentence to correct it.

1. A potential difference exists across the poles of a cell only when it is connected to an external circuit.

2. In a circuit, the sum of the voltage drops across the loads is only equal to the sum of the voltage rises across the voltage sources, if proper attention is paid to polarity when adding.

3. A voltage rise occurs when we trace a circuit through a voltage source from its positive pole to its negative pole.

4. The more a chemical cell is used, the more its potential difference will increase.

5. A primary cell is so called because it may be recharged.

6. George Leclanche pioneered the development of the wet cell in 1868.

7. The end voltage of a chemical cell is the voltage measured across the ends of the cell.

8. Carbon zinc flashlight cells should not be relied upon at temperatures below 32 °F (0 °C).

9. A dry cell is so called because it contains no water.

10. The alkaline cell uses a chemical system made up of manganese dioxide, water, zinc chloride, and zinc.

11. Mercury cells provide better low temperature service capacity than Leclanche or zinc chloride cells.

12. Magnesium cells are similar to Leclanche cells except for the substitution of magnesium for zinc in the negative pole.

13. Lithium cells produce a potential difference of about 1.35 volts when new.

14. Secondary cells are capable of being recharged by a charging potential which forces current through the cell from its negative pole to its positive pole.

15. A vented lead-acid cell is sometimes called a Plante cell after its inventor.

16. Sealed lead-acid cells come in a large range of sizes and service capacities from button to large rectangular cells.

17. A float charging potential of about 2.6 volts will bring a sealed lead-acid cell up to full charge within 24 hours.

18. A nickel cadmium cell uses a negative pole of nickelic hydroxide and a positive pole of cadmium.

19. Ni-cad cells show poor low-temperature performance.

20. Thermal runaway refers to the inability of ni-cad cells to deliver full power under high temperature conditions.

B. Briefly define the following terms:

1. Thermal runaway
2. Service capacity
3. Electrolyte
4. Float voltage
5. Local action
6. End or end-of-life voltage
7. Polarization
8. Characteristic potential difference of a chemical system (cite at least 4 examples)
9. The memory effect
10. Trickle charging

C. Experiments and Demonstrations:

1. Determining the service capacity of size AA Leclanche cells under varying conditions:

   a. Obtain 4 new AA-size Leclanche cells produced by any one of the well-known manufacturers and 4 size AA Leclanche cells produced by a "bargain" manufacturer or sold at a discount.

   b. Measure the open circuit potential differences of both standard and "bargain" cells using a digital voltmeter. Record your results on a chart like the one shown below.

Experiment 1 Part______	Brand A	Brand B
Initial potential difference	______volts	______volts
Potential difference at ____mA	______volts	______volts
After 10 minutes	______volts	______volts
After 20 minutes	______volts	______volts
etc.		

   c. Connect a 5-ohm, 1/2-watt or 1-watt resistor across one standard and one "bargain" cell. Measure the potential difference across each cell at ten minute intervals until both cells reach 0.9 volts. Record results on the chart. Which cell seems superior?

   d. Repeat part c using fresh cells and 150-ohm, 1/4-watt resistors (the initial current drain in this case is about 10 mA). Measure the voltage across each cell every 1/2 hour for 6 hours. Record your results on another chart.

   e. Having measured the cell potential differences, calculate the *average* current flowing through each cell during each measurement period. Calculate the service capacity by multiplying the currents by the time between measurements and adding all the products. Which cell shows a better service capacity under high load conditions? Under low load conditions?

f. Repeat part c using fresh cells that have been kept in the refrigerator freezer compartment for a few minutes. Return the cells to the freezer compartment between voltage measurements. Calculate the service capacity of the cells under low temperature conditions using the method of step e. Is there any difference in low temperature capacity between the cells?

g. Using the last pair of cells, repeat step c under the same low-temperature conditions used in step f. Again calculate the service capacities of the two cells. Is there a difference between the cells? (Low quality cells will probably show a greater loss of potential under low-temperature, low-current conditions than will cells of good quality).

h. Calculate the *cost* of an ampere-hour of service as supplied under various conditions from each of the cells

$$\frac{\text{cost of cell}}{\text{calculated a-h service capacity.}}$$

Is the "bargain"cell really a bargain?

2. Repeat experiment 1 using size AA Leclanche, zinc chloride, mercury, and fully charged ni-cad cells. Which cell gives the lowest cost per ampere-hour of service in each condition?

3. Repeat experiment 1 using lithium cells (be sure to recalculate the resistance required to give 300 mA and 10 mA currents using the rated voltage of the lithium cells you are using). How does the lithium cell compare with the other types in service and cost?

Glossary

This glossary is provided to give a ready reference to the meaning of some of the words with which you may not be familiar. These definitions may differ from those of standard dictionaries, but are more in line with standard shop usage.

ATR racking system A widely accepted size and mounting standard for airborne equipment.

alkaline cell A chemical cell using powdered manganese dioxide and zinc for its pole pieces and a water solution of potassium hydroxide for its electrolyte.

alternating current (ac) source A source of electric current and/or potential difference with two output terminals each of which regularly changes from being the positive pole to being the negative pole.

American Wire Gage (AWG) A wire sizing convention based on the circular mil as a measure of area. AWG size numbers are chosen so that subtracting three from the AWG number of any particular size wire will give the AWG number of the wire size with twice the cross section area and half the resistance per unit length.

amorphous or non-crystalline Said of a solid material in which the atoms are not arranged in a regular manner.

anode The electrode or conductor from which electrons are removed by an external circuit.

atom Originally thought of as a tiny, indivisible building block of matter. Late nineteenth and early twentieth century scientists found the atom to be composed of a nucleus, containing still smaller particles called protons and neutrons, and a number of electrons revolving on specific paths around the nucleus.

avionics A word formed from *avi*ation and electr*onics*. It refers to the use of electronics in the field of aviation in the broadest sense, including airborne, ground, and test equipment.

battery A connection or stacking of two or more cells to obtain more current or more potential difference or both than can be obtained from a single cell.

bleeder resistor The resistor of a voltage divider through which the smallest current flows. This resistor is generally chosen so that the current through it is about 10% of the total circuit current.

block diagram A kind of simplified schematic, that substitutes labelled boxes for functional sub-units and shows the generalized paths of controls and signals between the boxes.

cable 1) In common usage, any heavy conductor; 2) In electronics, two or more conductive paths bound into a single package.

carbon composition resistor The most widely used electronic component today. It is a resistor formed by embedding wire leads in a cylindrical slug of carbon and filler material. The hole is then usually covered with an epoxy or other plastic insulating jacket.

cathode The electrode or conductor to which electrons are added by a circuit.

characteristic potential difference The theoretical potential difference produced by a chemical cell using specific pole materials.

charge, or electric charge 1) An electrical unbalance in an object that is brought about by adding or removing electrons; 2) The physical condition that gives rise to an electric field.

charging current A current passed through a secondary cell for the purpose of recharging it.

chassis An aluminum, copper, or plated steel body around which an electronic unit is built. It serves as the support for the electronic components, and is often a voltage reference point.

chemical system The energy-producing mechanism of a particular chemical cell as defined by its pole materials and electrolyte.

circuit An entire complex of conducting paths and components.

circular mil By definition, an area equal to the area of a circle whose diameter is 1 mil.

$$A_{circular\ mils} = d_{mils}^2$$

closed circuit A continuous path from the negative pole of a cell or other source where there is a surplus of electrons, to the positive pole, where there is a deficiency.

coaxial cable (coax) A two conductor cable generally consisting of an outer plastic layer, a woven wire braid outer conductor, plastic insulator, and inner wire conductor. Coax is widely used for high frequency signal transmission.

color coding The use of color dots, bands, or stripes to identify wirc functions or the characteristics of electronic components.

conductor A material which permits the passage of an electric current. At normal temperatures, a metallic conductor contains many conduction electrons which are free to drift under the influence of an electric field.

constant current charging A method of recharging secondary cells by forcing a constant current through them in the direction opposite to that of normal current flow.

covalent bond A bond between two atoms which comes about when valence electrons are "shared" by the atoms.

crystal lattice The basic pattern in which atoms are arranged in a crystalline solid.

crystalline solid A material in which the atoms are arranged in a specific manner which is repeated throughout the solid.

current, conventional May be considered as the flow through a conductor of positive charge carriers from higher potential (positive pole) to a lower potential (negative pole).

current, electron flow The drift of conduction electrons through a conducting, closed circuit under the influence of an electric field. Electron motion is from the negative pole of a cell or source toward the positive pole.

depolarization The absorbtion of generated gases in a chemical cell, especially during "rest" periods. This may cause an apparent "rejuvenation" of the cell.

dielectric properties Refers to the "unbalanced" condition of the water molecule; one side of it is positively charged, the other side is negatively charged. Because of this, many ionic bond compounds are pulled apart in a water solution.

direct current (dc) source A source of electric current with two output terminals, one of which is always positive, the other always being negative.

discharge rate of a chemical cell The amount of current that a cell may be expected to furnish to an external circuit for a specific time, usually ten or fifty hours.

dissipation constant of a thermistor Symbolized by δ, a lower case Greek delta, the dissipation constant of a thermistor gives the ratio between the amount of current passing through the thermistor and the temperature increase due to the heating effect of this current. Usually given in milliwatts per degree centigrade.

distributed load A distributed load is one in which the opposition to the flow of an electric current is not confined to a particular point along the circuit. A long, high-resistance wire is an example of a distributed load.

dry cell A chemical cell in which the electrolyte is immobilized by absorption.

dyne The basic unit of force in the cgs system. It is equal to .00001 newton.

electric field 1) The space around a charged object; 2) The way in which space is modified around a charged object.

electron The negatively charged constituent of the atoms. Electrons may be found revolving around the atomic nucleus in specific paths called energy levels. Free or conduction electrons are electrons that have escaped from the attraction of their atomic nucleus and are free to drift through the material of which they are a part.

electron shell A grouping of electron paths or energy levels. Beginning with the shell closest to the nucleus, the shells are labelled K, L, M, N, etc.

electrostatics, or static electricity The phenomena associated with charged bodies. The phenomena associated with current flow are excluded from this classification.

electrovalent or ionic bond The bond formed by two atoms when one of them gives up one or more valence electrons to the other. The bond is based on the attraction between the positive and negative ions thus formed.

element An element is a simple substance formed of atoms of one type only. That is, all the atoms possess an identical or nearly identical nucleus and, when un-ionized, the same number of valence electrons.

end voltage (or end-of-life voltage) The voltage across a chemical cell when the cell should be discarded or recharged. The end voltage of a particular type of cell is usually defined by the manufacturer.

energy level A particular path that an electron may follow while revolving around an atomic nucleus. Each energy level may contain only up to a maximum number of electrons.

equivalent circuit A circuit containing only one or two components which has the same properties as a more complex arrangement and can be substituted for the more complex circuit for the purposes of analysis.

film resistor A resistor formed by coating a ceramic, glass, or other insulating cylinder with a metal oxide or other thin resistive film.

float charging potential A charging potential which may be left connected across the poles of a chemical cell on standby service without damage or destructive overcharge.

glow discharge tube A glass tube filled with a gas, such as neon, under low pressure. Two electrodes are embedded in opposite ends of the tube. When a sufficiently high potential difference is applied between the electrodes the gas will ionize and glow.

ground A reference point, often connected to a buried conductor in ground-side equipment, which is considered to be the 0 volt reference point. All voltages in a circuit are usually measured with reference to ground.

hole The absence of an electron at a point where one might be expected. For most purposes, a hole may be treated as a positive charge.

horse power 1) The amount of power required to lift 550 pounds a distance of one foot in one second. 2) In electrical equivalents:

$$1 \text{ hp} = 746 \text{ watts}$$

insulator For the purposes of this book, a material with a very high resistance to current flow. Few or no conduction electrons are present in an insulator.

ion An atom in which the negative charge of the electrons is not exactly balanced by the positive charge of the protons in the nucleus. An ion is formed when valence electrons are gained or lost by an atom.

Leclanche' cell (or carbon-zinc cell) A chemical cell using powdered manganese dioxide and zinc as its pole pieces and an ammonium chloride solution as its electrolyte.

Leyden jar A primitive capacitor. In effect, an apparatus for storing an electric charge on the inside foil lining of a glass jar.

linear resistance curve The characteristic illustrated by a load when any increase or decrease in the voltage across the load results in a proportional change in the current through the load.

lithium cell One of a family of chemical cell types incorporating lithium in one pole piece.

load An energy-absorbing or energy-using device of any sort, connected to a current source.

local action The formation of tiny chemical cells in one or both of the poles of a chemical cell due to impurities in the material. Local action may lead to exhausting of the service capacity of a cell or corrosion of the pole pieces.

logarithmic or audio taper potentiometer A volume control potentiometer whose resistance increases logarithmically as the control shaft is rotated in a clockwise direction.

lumped load 1) The theoretical load obtained by assuming that the resistance of a distributed load has been replaced by a single resistor or load. 2) A load whose resistance is so much greater than the distributed resistance in the circuit, such as that of the connecting wires, etc., that the circuit load may be considered concentrated at that point.

memory effect A reduction in the service capacity of ni-cad cells which occurs when cells on standby service are regularly discharged to only a small fraction of their full service capacity.

mercuric oxide or mercury cell A chemical cell using powdered mercuric oxide and powdered zinc as its pole pieces. The electrolyte is a water solution of potassium hydroxide.

mil 0.001 inch.

negative charge 1) The electrical charge possessed by the electron; 2) The charge on an object that results from the addition of electrons; 3) In Benjamin Franklin's conception, the result of removing electric "fluid" from an object.

neutron The uncharged constituent of the atomic nucleus. Possessing about the same weight as protons, neutrons seem to act as a sort of glue to hold the atomic nucleus together. Except for hydrogen, atomic nucleii contain at least as many neutrons as protons and in most cases there are more neutrons than protons.

Newton The basic unit of force in the mKs system.

nickel-cadmium cell A secondary cell utilizing pole plates made of nickelic hydroxide and metallic cadmium. A water solution of potassium hydroxide serves as the electrolyte.

nominal resistance of a thermistor The true resistance of a thermistor at a particular reference temperature. Most manufacturers use 20 °C as their reference temperature.

nominal value A marked or expected value which may be somewhat different from the actual measured value.

nucleus The relatively massive central portion of an atom. It is made up of a number of positively charged bodies called protons, and, except in the case of hydrogen, whose atomic nucleus usually contains only one proton, a number of uncharged bodies called neutrons.

Ohm's law Formula describing the relationship between the voltage applied to a linear load and the current through it:

$$R_{ohms} = \frac{E_{volts}}{I_{amperes}}$$

$$E_{volts} = R_{ohms} \times I_{amperes}$$

$$I_{amperes} = \frac{E_{volts}}{R_{ohms}}$$

open circuit A circuit which does not provide a continuous conduction path for electrons.

PVC (polyvinyl chloride) A popular, low cost, wire insulating material.

parallel connection of chemical cells All the positive poles are connected to one wire and all the negative poles are connected to another wire. (The current capacity of the parallel cells improves, the voltage across the battery remains the same as across a single cell.)

pictorial diagram The simplest kind of diagram used in electronics, it may be either a line drawing, sometimes with shading to emphasize shapes, a parts blow-up, or even a photograph of a piece of equipment, which provides us with information concerning the overall appearance of a unit, the shapes, relative sizes and location of components, interconnecting wires and cables, and so forth.

Plante' cell A secondary cell in which the pole pieces are formed of sheets of lead and lead dioxide. The electrolyte is a dilute solution of sulfuric acid.

point-to-point wiring Previously, the universally used method of building electronic units. Electronic components were mounted directly on the chassis and interconnected by means of wires that were integral parts of the components (leads), or by means of insulated hookup wire.

polarization A degradation in chemical cell performance, particularly in the case of Leclanche' cells caused by gas formation and the resulting insulation of portions of the pole area.

positive charge 1) The electrical charge possessed by a proton; 2) The charge on an object that results from the removal of electrons; 3) In Benjamin Franklin's conception, the result of adding electric "fluid" to an object.

potential difference The difference in the ability to do work possessed by a test charge (usually an electron or electron sized positive charge) at two different points. The potential difference between two points is measured in volts.

potential energy Technically speaking, potential energy is the result of a force operating through a distance. More commonly, it is the ability to do work possessed by an object if it is free to move in a gravitational or electric field.

potentiometer A three-terminal variable resistor, often used as a voltage divider.

power dissipation The power in watts turned into heat, light, or some other form of energy when an electric current passes through a component.

primary cell A chemical cell which is not considered to be rechargeable and is used until its service capacity is exhausted, after which it is discarded.

printed circuit board (PCB) Modern replacement for the electronic component chassis consisting of a plastic, fiberglass, or other insulating board with bonded copper strips for component interconnection. The components are generally mounted directly on the board by means of solder joining.

proton A positively charged building block found in the nucleus of the atom. The proton has roughly the same weight as the neutron, and is about 1850 times heavier than an electron.

pull-up resistor A resistor used to limit the current through a two-state device when the device is in its low resistance state, and to develop a potential difference when a two-state device is in its high resistance state.

quantum A minimal packet of energy.

reference designator Generally consisting of a combination of letters and numbers, the reference designator is used to uniquely identify a component or assembly. Reference designators are stackable, that is, a full reference designator identifies the major assembly, subassemblies, and, finally, the component.

relay rack (open or closed) A standard support system for ground equipment permitting the mounting of units with standard 19 inch wide panels. The open rack consists of two upright supports. The closed rack is actually a free-standing cabinet.

resistance Circuit elements that impede the flow of electrons.

resistivity (ρ[rho] in the MKS system) The actual resistance measured between a pair of opposite faces of a cube of material whose sides are one meter long.

resistor power dissipation rating The amount of power that a resistor may safely dissipate in the form of heat under controlled conditions.

resolution The ability to set and then reset a variable resistor to a specific resistance.

rheostat A two-terminal device rather like a potentiometer with one resistance element contact left out. The rheostat is generally a fairly high current device.

schematic diagram An arrangement of standardized graphic symbols for the functions of the parts of a piece of equipment. As long as an electrically conductive path exists between two points, a line is drawn between them; these lines may or may not represent actual wires. No indication is given of physical size or shape of units.

secondary cell A chemical cell which may be used until its service capacity is exhausted and then recharged to a state resembling its new condition.

semiconductor A class of materials which contain few free electrons at room temperature; but, because of the closeness of the valence and conduction energy levels, the addition of a small amount of energy is sufficient to raise many electrons to the conduction level.

series aiding connection of chemical cells Where the positive pole of one cell is connected to the negative pole of another. The potential (voltage) across the two will equal the sum of the potentials of the two cells separately. The current capacity remains the same as the current capacity of a single cell.

series-parallel arrangement Generally not distinguished from a parallel-series arrangement, refers to a complex arrangement of components involving both series and parallel combinations.

service capacity A measure of the amount of electrical energy that may be obtained from a chemical cell. Measured under controlled conditions and given in ampere-hours.

shunt resistor A resistor placed in parallel with some other component to reduce the current through that component.

statcoulomb The amount of charge on each of two bodies one centimeter apart which causes them to exert a force of one *dyne* on each other. The statcoulomb is the charge resulting from the addition of approximately 2×10^9 electrons to a body.

superconduction The effect experienced at very low temperatures when atomic vibration ceases and conduction electrons are free to drift through conductors with no opposition and no loss of energy.

temperature coefficient of resistance Symbolized by the Greek letter alpha (α), it is defined as the rate of change in resistance per degree centigrade of temperature change.

thermal runaway A destructive condition caused by overheating, especially in ni-cad cells, during charging at charge rates which are too high.

thermistor A resistor whose resistance varies with temperature in a precise, controlled manner.

time constant of a thermistor Symbolized by the Greek letter tau (τ), the time constant of a thermistor is defined as the amount of time it takes for a particular thermistor to change its temperature from some initial figure to 63% of the temperature of the air surrounding it.

tolerance The percentage variation in a characteristic of an electronic component.

transition element One of a series of elements which are consecutive in the periodic table and which possess similar chemical properties. All transition elements of a particular series possess the same number of valence electrons.

trickle charging A constant current charging method that keeps cells on standby service at full charge by passing a small current through them until they are put into service.

trimmer A potentiometer, generally with a screwdriver adjusted slider used for peaking or fine tuning a circuit.

two-state device An electronic component which may be switched to a high or low resistance state by a control signal.

valence electron An outer electron which may be given up or gained in the process of ionic compound formation, or an outer electron which is shared with another atom in the process of covalent compound formation.

variable resistor A resistor whose resistance may be varied by rotating an adjusting screw or shaft, or adjusting the position of a sliding contact.

vector An entity possessing both magnitude and direction, such as a force.

voltage divider A series of resistors placed across the poles of a source to provide a number of different voltages.

voltage drop The loss in potential energy in a circuit when a current is made to flow through a load and some of the energy is converted from electrical energy to some other form such as heat.

voltage dropping resistor A resistor placed in series with some other component in order to reduce the terminal voltage across or limit the current through that component.

voltage rise The increase in potential energy caused by a source such as a chemical cell connected in series aiding with the general current direction in the circuit.

voltaic cell Sometimes called a galvanic cell. It consists of two dissimilar pole pieces separated by an electrolyte, i.e., ionic conducting solution.

watt One watt is the power or work expended when a potential difference of one volt moves a current of one ampere through a circuit.

wet cell A chemical cell in which the electrolyte is in liquid form.

wire A single conductive path made of one solid filament or many fine metal threads twisted or braided together.

wire harness A cable formed of a number of insulated wires laced together.

wire-wound resistor A resistor manufactured by winding a high-resistance wire around a ceramic core or form.

wiring (interconnect diagram) Describes the electrical interconnections of the various components that go to make up a piece of equipment.

zinc chloride cell A chemical cell using powdered manganese dioxide and zinc as its pole pieces and a solution of zinc chloride as its electrolyte.

Answers to Self-Test Questions, Hints for Discussion Questions

Section 1

A. Self-Test Questions

1. The negatively charged particle found in the atom is called the *electron*. (Other negatively charged particles, such as *mesons* are produced when the atomic nucleus is split, but these particles have no role in electronics.)

2. The *proton* has a positive charge and is found in the nucleus of an atom.

3. An uncharged particle found in the nucleus of an atom is the *neutron*; it serves *to hold the atomic nucleus together*.

4. An outer electron which an atom may lose or gain in the process of forming a compound is called a *valence* electron.

5. An electrically charged body is surrounded by an *electric field*.

6. When an atom gives up an electron to another atom it becomes an *ion*. When an atom shares a pair of electrons with another atom, it forms a *covalent bond*.

7. An electrode to which electrons have been added is called a *negative electrode*; it has a negative charge.

8. Conduction of electricity in a gas takes place through the movement of *ions and electrons* under the influence of an *electric field*.

9. Conduction of electricity in a semiconductor takes place through the movement of *electrons* and *holes* under the influence of an electric field.

10. A quantum is a minimal *packet of energy*.

11. a. When an electron in a particular electron shell absorbs energy, it moves to a higher energy level within that shell. If the electron absorbs enough energy, it will move to a shell further from the atomic nucleus, or it may leave the nucleus completely and become a *free* or *conduction electron*.

    b. When a free or conduction electron absorbs energy its velocity of motion is increased.

12. a. Your answer should include the following basic notions:

    1. An element is a basic substance which is formed of atoms which are identical or nearly identical in weight, structure, and in the ways in which they combine with the atoms of other elements;

    2. Elements are the building blocks that are combined to produce all the matter in the universe;

    3. When not ionized, all atoms of a given element contain the same number of electrons.

    b. At present, 103 elements have been either discovered in nature or manufactured in the laboratory. Several others may be produced as a result of high energy nuclear experiments as time passes.

13. An electron shell consists of a number of distinct paths upon which electrons may be found around an atomic nucleus. The shells may contain from one path and two electrons, as in the case of the K shell, to as many as 36 paths and 72 electrons in the case of the P shell.

14. A valence electron is an outer electron which may be transferred from, or to, an atom in

the process of ionic compound formation. When an atom has donated or accepted a valence electron, it is no longer electrically neutral; it has become a positive or negative ion. Conduction electrons are found in electrical conductors such as copper. Conduction electrons may be said to "belong" to the mass of material as a whole, rather than to individual atoms. This is the case when the energy level of the outermost electron(s) of an atom is sufficient to put the electron(s) in a conduction band. Some common materials in which we find conduction electrons are copper, silver, gold, iron, and nickel.

15. The crystalline structure of copper is the face-centered cubic arrangement. Germanium generally exhibits a diamond or tetrahedron structure.

16. The principal differences among conductors, semiconductors, and insulators derive from the presence or absence of conduction electrons. Few or no conduction electrons are found in insulating materials. In a conductor, on the other hand, most of the valence electrons form an electron cloud of conduction electrons in the material. In a semiconductor, the energy levels of valence and conduction electrons are so close that valence electrons may be raised to the conduction band level by absorbing energy from an energy source such as an electric field.

B. Questions for Thought

1. An atomic reactor (b) derives its ability to do work by using the energy stored in the atomic nucleus. All the other "devices" indirectly draw their energy from the sun, which is itself a sort of atomic reactor.

2. Since the atomic nucleus is *not* involved in the passage of an electric current, a knowledge of the structure of a particular nucleus can be of use only in helping to determine a) the sort of ion that the atom may form; or, b) how strongly the valence electrons will be bound to the atom. In any event, the crystalline structure of an element is of more importance.

3. Lithium, sodium, and potassium all belong to group I of the periodic table, forming $+1$ ions. Radon, argon and xenon belong to group VIII, and therefore do not generally form ions at all. Bromine, chlorine, and fluorine belong to group VII, forming -1 ions. Sulfur and tellurium belong to group VI, generally forming -2 ions, although other ionic forms are possible.

Section 2

A. Self-Test Questions

1. Contact between a woolen cloth and a piece of amber or plastic comb will transfer *electrons* to the amber or comb, resulting in a *negative* charge on the amber or comb and a *positive* charge on the cloth.

2. A charged body may be used to charge a neutral object some distance away from it by a process known as *induction.*

3. Late nineteenth century scientists thought that the *electromagnetic ether* (or aether) transmitted light, magnetic, and electrical forces between two bodies not in contact.

4. Faraday's method of charting an electric field by means of lines of force is capable of showing the *strength* (or *magnitude*) and the *direction* of the field as well as the direction a small *positive* (or *positive test*) charge would move if placed at a point in the field.

5. A force is a *vector* quantity; to symbolize this we can draw a small *arrow* above the letter representing the force.

6. Work may be defined as a *force* acting through a *distance.*

   Mathematically this is symbolized as

$$W = \vec{F} \times d$$

7. The force is equal to

$$\frac{q_A \times q_B}{d^2} \quad \text{or} \quad \frac{1}{16} = 0.0625 \text{ dyne.}$$

8. The potential is equal to

$$\frac{q}{d} \text{ or } \frac{1}{4} = 0.25 \text{ statvolts.}$$

9. A voltaic cell is composed of:

   a. a positive pole

   b. a negative pole

   c. an electrolyte solution.

10. One volt is equal to *0.0033 (or 1/300)* statvolt; one coulomb is equal to 3×10^2 catatcoulombs.

11. Your answer here should point out that both gravity and electricity involve forces acting over a distance, and both involve fields and potentials.

12. A lightning rod protects a house by providing a discharge path to ground for static charges in the nearby clouds.

13. Potential energy is the ability to do work possessed by an object in a field. The potential energy possessed by the object is equal to the work done against the field when the object is moved to its place.

14. The inverse square law states that two quantities A and B are related by a formula of the form

$$A = \text{constant} \times \frac{1}{B^2}$$

(where B is often a distance).

Electrostatic force and gravitational force are examples of the inverse square law. Potential is not.

15. The basic units of the mks system are the meter, the kilogram, and the second. The basic units of the cgs system are the centimeter, the gram, and the second.

16. Briefly, the voltaic cell operates by transferring electrons from its positive pole to its negative pole. The chemical reactions found in commercial cells are discussed in greater detail in Section 9.

B. Questions for Thought:

1. Your answer might include dust collection and atom smashing.

2. Unfortunately, the switch from the English to the metric system is becoming a political and emotional issue. Some of the advantages of the metric system are the simplicity of many formulas, its increased ease of use, and its wide acceptance. The prime disadvantage is the great cost required to replace scales, tape measures, tools, etc., with metric equivalents.

3. Increasing the diameter of the pole pieces and spacers will increase the current available from the cell.

4. Self-explanatory

Section 3

A. Self-Test Questions

1. A DC source has *two* output terminals, one of which is always the *positive* terminal, the other is always the *negative* terminal.

2. A *pictorial* diagram shows the relative sizes, shapes, and locations of components in a unit.

3. A *reference designator* is a combination of letters and numbers which is used to uniquely identify an assembly or a component.

4. The *schematic* diagram conveys the most information about the function of the various components; it is made up of graphic *symbols*.

5. The symbol ⊣|ı⊢ represents a battery.

6. An *open* circuit may be used to transfer a potential difference.

7. A *closed* circuit permits the flow of *current.*

8. A current does not exist in *an open* circuit.

9. An increase in the current capacity over that of a single cell is obtained in a *parallel* arrangement; an increase in potential difference is obtained from a *series* circuit.

10. Two equal cells arranged in a *series bucking* arrangement produce a potential difference of zero across the battery thus formed.

B. Questions for Thought

1 a. Which sort of diagram would show the assembler how to arrange component leads?

b. The engineer is working with ideas--the unit isn't ready to build yet.

c. Does he have to know what is in the radio? If not, what sort of diagram might he use?

d. In this case, can we eliminate any sort of diagram?

Section 4

A. Self-Test Questions

1. One *milli*ampere is equivalent to 0.001 ampere.

2. The symbol Ω is the usual abbreviation for *ohm(s).*

3. Ten thousand herz is normally written as ten *kilo*herz (or 10 KHz).

4. 1 MΩ is short for one million ohms.

5. One nanofarad = 1000 picofarads = 0.001 microfarads.

6. Micromicrofarad, abbreviated mmf, is the older form of *picofarad.*

7. Ten gigaherz (GHz) $= 1 \times 10^4$ megaherz ($1 \text{ MHz} = 1 \times 10^6$Hz).

8. 18 milliampere (18 mA) $= 1.8 = 10^4$ nanoamperes (nA)$= 1.8 \times 10^7$ microamperes (μ).

9. 33.6 μ volts = 0.0000336 volt.

10. One microvolt $= 1 \times 10^{-9}$ kilovolts.

11. Twelve picofarads = 0.012 nanofarad.

12. Thirty centimeters (cm) = 300 millimeters (mm) = 0.3 meter (m).

13. 6.66 milliamperes or 0.0066 ampere.

14. 8.18 mA, 4.09 volts.

15. 454.54 ohms.

B. Questions for Thought

1. a. Total current is 0.15 A, voltage drop across load A is 1.5 volts.

2. Self-explanatory

3. Nonlinear. (Does the current through the circuit depend on the voltage?)

4. One is easy to find; there are many others.

Section 5

A. Self-Test Questions

1. If two 80-ohm loads are connected in series the resulting resistance is *160* ohms.

2. Two 5-ohm loads in parallel may be replaced by an *equivalent* load of 2.5 ohms.

3. Adding another load in series with two other series-connected loads in a circuit will *decrease* the circuit current and *reduce* the voltage drop across each load. The resistance of the entire circuit will *increase*.

4. Adding another load in parallel with two other parallel loads in a circuit will *increase* the circuit current. The resistance of the entire circuit will *decrease*, and the potential drop across each load will *remain the same*.

5. In a parallel circuit composed of two loads the total resistance of the parallel combination:

$$R_t = \frac{R_A \times R_B}{R_A + R_B} \quad \text{or} \quad \frac{1}{\frac{1}{R_A} + \frac{1}{R_B}}$$

6. 99.6 or 100 watts.

7. About 40 mA.

8. 0.288 watt.

9. 28.8 KΩ.

10. About 62.5 mA (from a 120-volt line).

B. Problems for Calculation

1. 12.5 Ω.

2. The voltage drop across each load is 12 volts. The current through the 180Ω load is 66.6 mA, 80 mA through the 150Ω resistor, and 100 mA through the 120Ω resistor.

3. $R_X = 30$

4. $I = 0.255$ A, $V_{RX} = 1.275$ volts.

5. Current drawn is 9.58 A. The heater could *not* be connected. An additional 650.4 watts could be drawn without opening the breaker.

Section 6

A. Self-Test Questions

1. Most *ground side* equipment is designed to permit mounting in standard *relay racks* which require a panel width of 19 inches (nominal width).

2. The *ATR* racking system is used for most airborne equipment.

3. The aluminum or plated steel body upon which electronic equipment is built is called a *chassis*.

4. The method of construction in which the components are connected by means of their leads or by separately connected hook-up wires is called *point-to-point* wiring.

5. A glass fiber or plastic board with bonded copper strips for electrical connection is called a *printed circuit board*.

6. A *black* hook-up wire might mark a ground connection in a vacuum tube or transistor circuit, while a red wire would probably be used for *positive high voltage* or the *B+ supply line*.

7. a. 1 KΩ, 5%.

   b. 5.6Ω, 10%, 0.01%.

   c. 27KΩ, 5%, 0.001%.

   d. 10MΩ, 20%.

   e. 910Ω, 2%.

8. Using $A = d^2$, 265.69 circular mils.

B. Questions for Thought

1. Depending on its use, the outer braided shield may act as electrostatic shielding, a second current path, a structural strengthener, or all of these.

2. Your description should include the following points:

   a. The AWG system is primarily used in America to give the diameter-based gauge size of wire.

   b. A mil is 0.001 inch.

c. A circular mil is defined as an area equal to the area of a circle whose diameter is 1 mil.

d. $A_{(circular\ mils)} = [d_{(mils)}]^2$.

3. Using equation 6.4, $R_{75\,°C} = 65.58\Omega$.

4. Again using equation 6.4, $R_{80\,°C} = 9700\Omega$.

5. Using Table 6-6 and equation 6.4, $R_{75\,°C} = 19.55\Omega$.

6. Converting to meters and using equation 6.2, $R = 40.24\Omega$.

7. One point that should be made is the advantage to the airline companies of reducing the number of spare unit types required.

Section 7

A. *Self-Test Questions*

1. line a, 3900, 4290, 3510.

   line b, brown, black, blue, 11 MΩ, 9 MΩ

   line c, blue, gray, brown, 680.

   line d, 10,000, 10,500, 9,500.

   line e, 8.2, 8.61, 7.79.

   line f, yellow, violet, brown, 517, 423.

   line g, 10 MΩ, 11 MΩ, 9 MΩ.

   line h, green, orange, yellow, gold.

   line i, yellow, orange, brown, 438.6, 421.4.

   line j, 7.5 KΩ, 7875, 7125.

2. a. Resistor power dissipation may be determined from *the size* of the resistor for resistors with a power dissipation of 2 watts or less.

   b. A 2-watt, 5%, 100Ω resistor will dissipate *more* heat than a 0.1 watt, 10%, 1000Ω resistor.

   c. For high power dissipations, *wire-wound* resistors are most often used.

   d. *High cost* and *problems in handling high frequency AC* are two of the principal drawbacks in the use of wire-wound resistors.

   e. The most frequently found components in avionics circuits are *carbon composition resistors*.

   f. The resistance of a carbon composition resistor will *slowly change* throughout its life span.

   g. Carbon composition resistors must be treated carefully; they are sensitive to:

      1. exceeding rated power dissipation
      2. moisture
      3. the shock of temperature change during operation or construction.

   h. In high voltage or low noise applications *film* or *wire-wound* resistors should be used, *carbon composition* resistors should be avoided.

   i. A shunt resistor is placed in *parallel* with its load; it is primarily a *current carrying* device.

   j. A series resistor may be used to *reduce* voltage or limit *current*.

   k. Film resistors are available in power dissipation ratings from 1/20 watt to 2 watts.

   l. Carbon composition resistors *may not* be used to replace film or wire-wound types.

m. A burned out 10% tolerance resistor *may* be replaced by a 5% tolerance resistor of equal resistance, wattage, and reliability.

n. Connecting two 10% tolerance resistors in parallel *will not* provide the same precision as one 5% tolerance resistor.

o. A bleeder resistor is used in a *voltage divider* circuit; it passes *less* current than any other resistor in that circuit.

p. Two 100Ω, 1-watt resistors in parallel *may* be used where a 50-ohm, 2-watt load is required for test purposes.

q. Choice of a meter shunt resistance is based on the *full-scale current* and the *resistance* of the meter.

r. The choice of bleeder resistor current is based on *10% of the total current supplied by the voltage divider.* The value may be adjusted to make it possible to use standard value resistors in the divider.

s. If a shunt resistor burns up, creating an open circuit, the component with which it is in parallel may be destroyed by *the increased current through it.*

t. If a series voltage-dropping resistor opens, the load in series with it will *probably not* be destroyed.

u. The high-end range of an ammeter, that is, its ability to measure high current, can be increased by putting a *shunt resistor* in *parallel* with the meter.

B. Questions for Thought

1. Using a 10% current value for the bleeder, the theoretical values calculated are:

$R_1 = 9.7\Omega$, 23W

$R_2 = 4.85\Omega$, 8.7 W

$R_{b1} = 39.3\Omega$, .77W

A one-watt carbon composition may be used for the bleeder.

2. Connect the buzzer and motor in series, shunt the buzzer with a 3Ω resistor.

3. Cost should certainly be a prime criterion. On the other hand, you must stock those components that might be required frequently, in order to save time and trouble.

Section 8

A. Self-Test Questions

1. a. A thermistor possesses a *negative* temperature coefficient of resistance. Its resistance *decreases* as its temperature increases.

b. The resolution of a variable resistor is a measure of the ability *to select a specific resistance repeatably.*

c. The familiar front-panel *volume* control is most frequently a *logarithmic-* (or *audio-*) *taper potentiometer.*

d. Trimmer potentiometers are frequently used for *fine tuning a circuit.*

e. A rheostat is a *two-terminal* device; a potentiometer is a *three-terminal device.*

f. The power dissipation rating of most of the potentiometers used in avionics equipment is less than *4 watts.*

g. Carbon potentiometers are generally available with total resistances of *50Ω* to *15 MΩ.*

h. The time constant τ (tau) of a thermistor is defined as the time it takes for a thermistor to change its temperature from some initial temperature to 63° of the temperature of the air surrounding it.

2. δ dissipation constant

   α temperature coefficient of resistance

   τ time constant

   R_{20} nominal resistance

B. Questions for Thought

1 a. The circuit could be a series connection of a rheostat for dimming, a voltage dropping resistor that would prevent the voltage drop across the lamp from exceeding 8 volts even when the rheostat was set at zero, and the lamp itself.

   b. Self-explanatory

   c. The series dropping resistor would be 13.3 ohms (15 ohms as a practical value), and, allowing a safety factor, a power dissipation of 2 watts. The rheostat should have a maximum resistance of 20Ω and a power dissipation of 1.8 watts (say 2 watts, again allowing a safety factor).

2. If a resistor were substituted for the thermistor, the amount of voltage fed back through the phase shift to the summing point would be a *fixed* percentage of the operational amplifier output voltage. There would still be some automatic control over the amplifier output.

3 a. (Potentiometer in full CW position) voltage drop across load = 22.4 volts, current = 22.4 mA.

   b. (Potentiometer fully CCW) voltage drop across load = 2.18 volts, current = 2.18 mA.

   c. One practical application for this circuit would be to furnish a load with an input voltage that could be varied continuously through a range of ten times the lowest voltage, i.e., the range would be from 2.2 to 22 volts. The two 1.2 KΩ resistors define the maximum and minimum voltage applied to the load.

4. Full brightness control could not be obtained if one or two of the three bulbs burned out, unless the maximum resistance of the rheostat were increased greatly.

5. $R_{20°}$ could be measured directly. $R_{37°}$ could be measured after holding the thermistor in a closed fist for several minutes. The temperature coefficient of resistance, α can be determined if these two are known. Allowing the thermistor to cool to 20°C again and noting the time taken, we can determine τ, which would be the time taken to cool to 63% of the temperature difference between 37°C and 20°C. The dissipation constant may be determined by connecting the thermistor across the power supply and increasing the voltage while measuring the current through the thermistor. When the resistance of the thermistor, as calculated from the voltage and current, is the same as that when the thermistor was raised to body heat ($R_{37°}$), the power output of the power supply in millivolts divided by 17°C, the temperature rise, will be equal to δ, the dissipation constant.

Section 9

A. Self-Test Questions

1. False. A potential difference always exists across the poles of a cell, unless the electrolyte has become totally non-conducting or the poles are internally short-circuited.

2. True.

3. False. Using the conventional current method, a voltage rise occurs when we trace a circuit through a voltage source from its negative pole to its positive pole.

4. False. Only in the case of the lithium cell does the potential difference increase slightly at first. All other chemical cells

and lithium cells after the first few hours of use, show a decrease in cell potential difference as they are used.

5. False. A primary cell is generally not considered rechargeable. A secondary cell is.

6. False. Leclanche pioneered the *dry* cell.

7. False. The end voltage of a chemical cell is the potential difference across the cell when it is assumed to have reached the end of its useful life. The end voltage is usually a figure determined by the manufacturer.

8. True. Although they will still show some activity, carbon-zinc cells have a very much reduced service capacity below 32 °F (0 °C).

9. False. A dry cell is so called because the water in its electrolyte is absorbed in porous material and is not free to flow.

10. False. The alkaline cell uses a chemical system made up of manganese dioxide, zinc, water, and potassium hydroxide.

11. False. Mercury cells provide poorer low temperature service.

12. False. The electrolyte used in magnesium cells consists of bromine or magnesium, and chlorine.

13. False. Whatever the particular chemistry used in a lithium cell, its starting potential difference is 2 volts or more.

14. False. The charging potential forces a current through the cell from its positive to negative pole.

15. True.

16. False. Presently, sealed lead-acid cells come in very few sizes.

17. False. A float charging potential of between 2.3 and 2.4 volts will bring a sealed lead-acid cell up to full charge within 16 hours.

18. False. A ni-cad cell uses a positive pole of nickelic hydroxide and a negative pole of cadmium.

19. False. Ni-cad cells show fairly good low temperature performance.

20. False. Thermal runaway refers to the build-up of heat in a charging cell. This build-up of heat may lower the cell resistance causing a greater charging current and increased heating. Overheating or an explosion may result.

B. Definitions

1. See answer to question 20.

2. The service capacity of a cell is its ampere-hour rating, that is, the amount of current which a cell may be expected to furnish times the length of time that this current will be delivered.

3. The electrolyte is the ionic conductor solution that permits the transfer of electrons from the positive pole of the cell to the negative pole.

4. The float voltage is that voltage which may be connected across the poles of a secondary cell in such a way as to cause continuous charging, thereby keeping the cell at a 100% charge without causing cell damage.

5. Local action generally occurs in the negative pole of dry cells. It consists of the formation of a tiny short-circuited cell between two sections of the negative pole because of impurities in the pole metal. It causes holes to form in the pole piece. Local action is also responsible for the discharge of secondary cells which are being stored.

6. The end or end-of-life voltage is that voltage which exists across a primary cell when the cell must be replaced. It is also the voltage reached by a secondary cell when the cell must be recharged. The end-of-life voltage is normally specified by the cell manufacturer.

7. Polarization is most generally noted in carbon-zinc cells. It is the formation of tiny

gas bubbles on the positive pole particles which serve to insulate these particles and slow or stop the chemical action of the cell.

8. The characteristic potential difference of a chemical system is the potential difference determined by the atomic structure of the pole materials. It is independent of the structure and size of the cell, and, in general, of the electrolyte. Examples given in this text are:

   a. manganese dioxide-zinc, 1.55 volts

   b. mercuric oxide-zinc, 1.3 volts

   c. silver oxide-zinc, 1.6 volts

   d. lead dioxide-lead, 2.2 volts.

9. The memory effect is most often found in those nickel-cadmium cells which are used in standby service. This condition occurs when a cell is discharged a number of times to only a small percentage of its service capacity. Somehow, the cell *remembers* the level of its earlier discharges, and will not provide a greater service capacity than the *remembered* level.

10. Trickle charging is the method of keeping a cell on standby service at full charge by passing a small charging current equal to the 500-hour discharge rate through the cell.

DC Circuits Final Examination

STUDENT ____________________

GRADE ____________________

Place a circle around the letter indicating the correct answer.

1. Resistance measurements are made on two thin metallic rods. Rod A is a good conductor having a low resistance, rod B has a high resistance. When exposed to very strong light, the resistance of A increases, while the resistance of B decreases greatly. Using your knowledge of the Bohr model of the atom and crystal structure, explain these observations.

2. Three new Leclanche cells are connected between terminals A and B.

   A. Draw a schematic diagram showing how the cells are connected if the potential difference between A and B is 4.5 volts.

   B. Draw a schematic diagram to show how the three cells may be connected to give a potential difference of three volts.

   C. Draw two different schematics to show two connections that would produce a potential difference of 1.5 volts.

   D. Circle the arrangement which would produce the greatest current if terminals A and B were short circuited.

3. Fill in the blanks in the following sentences.

   A. Faraday's method of charting an electric field by means of lines of force is capable of showing the ____________ and ____________ of the field as well as the direction a small ____________ test charge would move if placed at a point in the field; the closer together the lines of force, the ____________ the field.

   B. A force is a ____________ quantity; to symbolize this we can draw a small ____________ above the letter representing the force.

   C. Work may be defined as a ____________ acting through a ____________. Mathematically, this is symbolized as:

   W = ____________ × ____________

   D. The force on a test charge of one statcoulomb four centimeters away from another charge of one statcoulomb is ____________ ____________.

   E. After removing the test charge used in part D, the potential at the point is ____________ ____________.

   F. If a statcoulomb is the charge carried by 2.08×10^9 electrons, and a coulomb is about equal to 3×10^9 statcoulombs, the number of electrons passing a point in a circuit with a current of 1 ampere in one second is ____________.

4. Give a working definition of a DC source.

5. What sort of information can be obtained from:

   A. A pictorial diagram.

   B. An interconnect or wiring diagram.

   C. A schematic diagram.

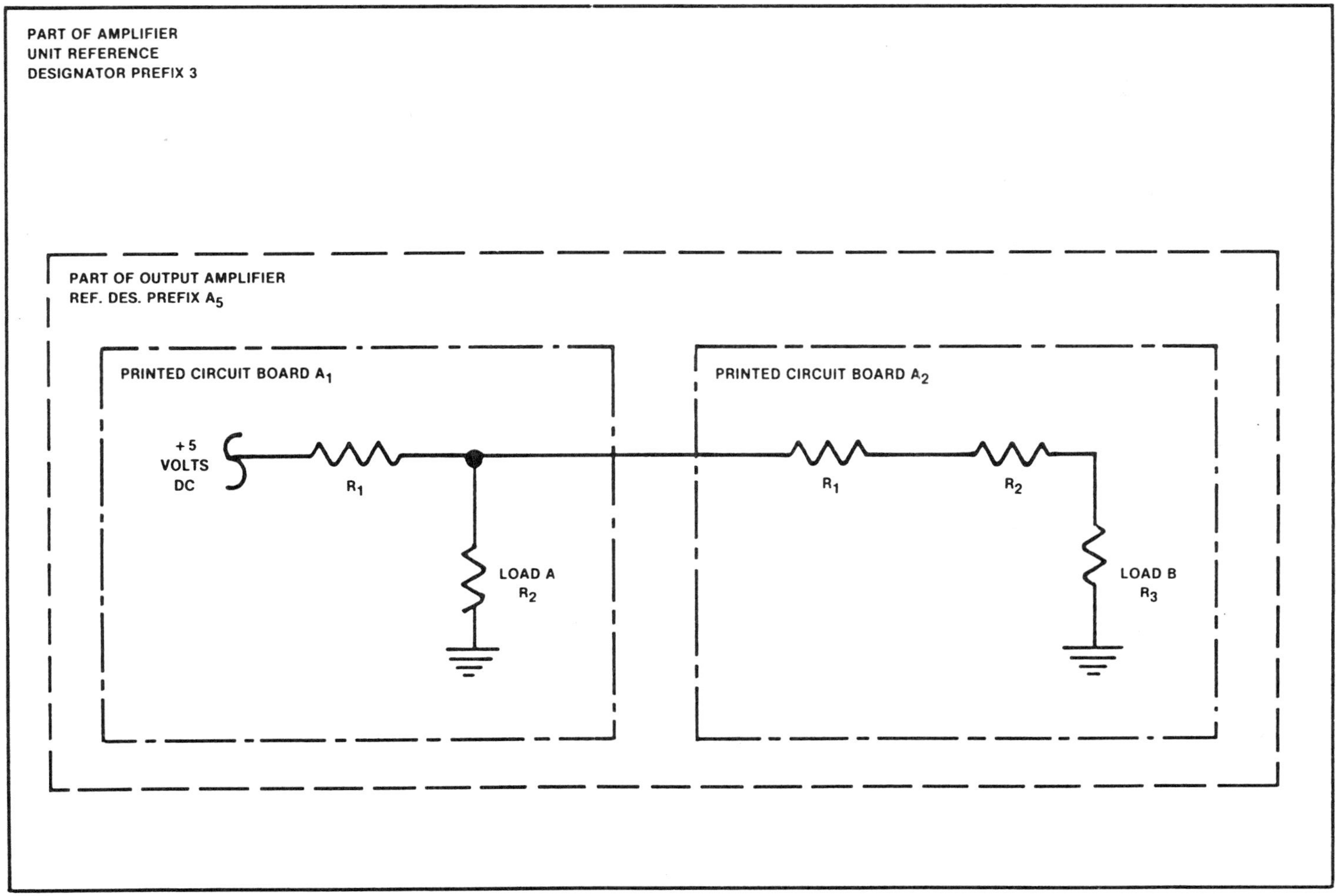

PART OF AMPLIFIER
UNIT REFERENCE
DESIGNATOR PREFIX 3

PART OF OUTPUT AMPLIFIER
REF. DES. PREFIX A_5

PRINTED CIRCUIT BOARD A_1

+5
VOLTS
DC

R_1

LOAD A
R_2

PRINTED CIRCUIT BOARD A_2

R_1

R_2

LOAD B
R_3

Fig. A1

NOTE: ALL RESISTORS ARE 100 Ω, 1/4W, 5%.

6. Figure A1 represents a typical schematic diagram used by avionics equipment manufacturers. Notice the "stacking" of reference designators.

   A. What is the complete reference designator for load A?

   B. What are the color code bands found on the resistor which serves as load B?

   C. What is the current through 3A5A2R2? (Assume nominal resistor values in this and following parts.)

   D. What is the current through load A?

   E. What is the voltage drop across load A?

   F. How much power is dissipated by load B?

7. Fill in the following table:

Prefix	Decimal Value	Exponential Value
a. milli	....	10^{-3}
b. pico	....	....
c.	0.000001	....
d.	1000	....
e.	....	10^{6}
f. nano	....	....

8. Using E for voltage, I for current, and R for resistance, state Ohm's law in three forms.

9. Give two formulas which may be used to calculate the equivalent resistance R_{eq} that may be substituted for a parallel combination to two resistors R_A and R_B.

10. Will the power dissipation rating of a 91-ohm 1/4 watt resistor be exceeded if it is connected across a 5-volt source? What will the power dissipation be?

11. If a 150-ohm resistive load is dissipating exactly 0.25 watt, the current through it is ______________ mA.

12. Complete the following sentences:

   A. The nominal panel width of a unit designed to be mounted in a standard relay rack is __________ in.

   B. The __________ racking system is frequently used for mounting airborne equipment.

   C. Color bands of grey, red, red, and red specify a resistor with a nominal resistance of ________ ohms and a tolerance of __________%.

   D. The color coding often used for the insulation of a wire at ground potential is __________; the color __________ generally signifies a B+ supply line.

   E. Although __________ is a better conductor of electricity than copper, its high cost limits its use in electronics.

13. Fill in the resistance value and tolerance symbolized by the following color bands:

   A. Brown, black, black, gold __________

   B. Green, blue, gold, gold __________

   C. Brown, brown, brown, brown __________

   D. Red, violet, orange, none __________

   E. Orange, white, green, silver __________

14. Complete the following sentences:

   A. A __________ is a two-terminal variable resistor frequently carrying a fairly high current.

   B. A __________ is a three-terminal device generally used to provide a specific voltage between its moving contact terminal and ground.

   C. In speaking of a thermistor α (alpha) is defined as ______________.

   D. The bleeder resistor in a voltage divider is generally calculated to pass a current roughly __________ __________.

15. Complete the following table:

Type of Cell	Chemical System	Open Circuit Voltage of New Cell	Low Temperature Performance poor/fair/good
Leclanche			
Alkaline			
Mercury			
Plante			
Ni-cad			

16. Briefly define the following terms:

A. Service capacity of a chemical cell.

B. Thermal runaway.

C. Float voltage.

D. Electrolyte.

E. Memory effect in a chemical cell.

Answers To Final Examination

DC Circuits

1. Rod A is a metallic conductor and has a relatively low resistance, that is, the drift of the conduction electrons is not seriously impeded by the cyrstal lattice. When exposed to strong light, however, the atoms in the lattice absorb energy and vibrate more strongly, increasing the resistance to conduction electron drift. Rod B is composed of a semiconducting material. Under normal conditions, most of the outer electrons remain in the outer energy level of each atom as valence electrons; few free or conduction electrons are present. However, in materials of this sort, the energy possessed by a conduction electron is only slightly higher than the energy of a valence electron, so that when a strong light is shone on the rod, a large number of valence electrons absorb quanta of energy and become conduction electrons. This results in a lowering of the resistance of the rod.

2. a. $V_{AB} = 4.5v$

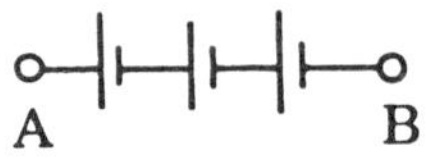

   b. $V_{AB} = 3v$

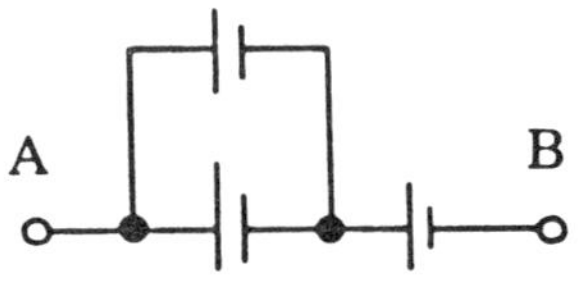

   c. $V_{AB} = 1.5v$

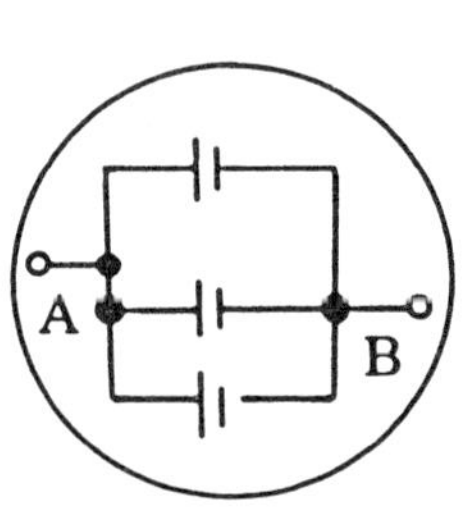

   or

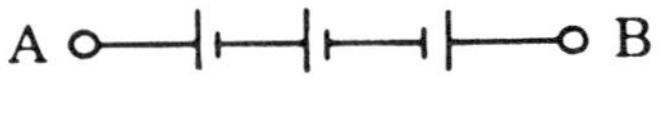

3. a. Faraday's method of charting an electric field by means of lines of force is capable of showing the *strength* (or *magnitude*) and *direction* of the field as well as the direction a small *positive* test charge would move if placed at a point in the field; the closer together the lines of force, the *stronger* the field.

   b. A force is a *vector* quantity; to symbolize this we can draw a small *arrow* above the letter representing the force.

   c. Work may be defined as a *force* acting through a *distance*. Mathematically, this is symbolized as:

$$W = \vec{F} \times d$$

   d. The force on a test charge of one statcoulomb four centimeters away from another charge of one statcoulomb is 0.0625 dyne.

   e. The potential at the point is *0.25 statvolts.*

   f. The number of electrons passing a point in a circuit with a current of 1 ampere in one second is about 6.24×10^{18}.

4. A dc source possesses two output poles or terminals across which there is a potential difference. One of the poles is always the positive pole, the other is always the negative pole.

5. a. A pictorial diagram provides information concerning the overall appearance of a unit, the shapes, relative sizes, and location of components, interconnecting wires and cables, etc. It also identifies

the components in such a way as to permit cross checking with a parts list. One special sort of pictorial, the assembly, or parts blow-up, provides information which may aid in the disassembly or reassembly of a unit.

b. The interconnect or wiring diagram is especially useful to those who must install or wire an electronic unit. It specifies the colors and wire sizes of the interconnecting wires, their points of connection and their layout.

c. The schematic diagram which uses standardized graphic symbols for component functions is arranged to provide the maximum amount of information about the function of the various components in a unit.

6. a. The complete reference designator for load A is 3A5A1R2.

b. Brown, black, brown, gold.

c. I = 28.5 mA.

d. I = 21.37 mA.

e. 2.14 volts.

f. 0.005 watt or 5 milliwatts.

7. a. milli-, 0.001, 10^{-3}.

b. pico-, 0.000000000001, 10^{-12}.

c. micro-, 0.000001, 10^{-6}.

d. kilo-, 1000, 10^{3}.

e. mega-, 1,000,000, 10^{6}.

f. nano-, 0.000000001, 10^{-9}.

8. a. I = E/R

b. E = IR

c. R = E/I

9. a.

$$R_{eq} = \frac{R_A \times R_B}{R_A + R_B}$$

b.

$$R_{eq} = \frac{1}{\frac{1}{R_A} + \frac{1}{R_B}}$$

10. Yes. The power dissipated will be 0.27 watt, slightly greater than the 0.25 watt dissipation rating of the resistor.

11. If a 150-ohm resistive load is dissipating 0.25 watt, the current through it is 4 mA.

12. a. 19 in.

b. ATR.

c. 8.2 kilohms (or 8200 ohms); 2%.

d. black; red.

e. silver.

13. a. 10Ω, 5%.

b. 5.6Ω, 5%.

c. 110Ω, 1%.

d. 2700Ω (or 27KΩ), 20%.

e. 3.9 Megohms, 10%.

14. a. Rheostat.

b. Potentiometer.

c. The temperature coefficient of resistance.

d. Ten percent of the total current furnished by the divider.

15.

Type of Cell	Chemical System	Open Circuit Voltage	Low Temperature Performance
Leclanche	Manganese dioxide, zinc, water solution of ammonium chloride	1.5	poor
Alkaline	Manganese dioxide, zinc, water solution of potassium hydroxide	1.5	good
Mercury	Mercuric oxide, zinc, water solution of potassium hydroxide	1.35	poor
Plante'	Lead dioxide, lead, water solution of sulfuric acid	2.2	good
Ni-cad	Nickelic hydroxide, cadmium, water solution of potassium hydroxide	1.2	good

16. a. The service capacity of a chemical cell is its ampere-hour rating, that is, the amount of current which a cell may be expected to furnish times the length of time that this current will be delivered.

b. Thermal runaway is a condition caused by the buildup of heat in a chemical cell, generally a ni-cad cell, which is being charged at too high a rate. The heating effect of the charging current may lower the cell resistance causing a greater charging current to flow. This in turn will cause an even greater buildup of heat and so on. Boiling of the electrolyte or an explosion may result.

c. The float voltage is that voltage which may be connected across the poles of a secondary cell in standby service in such a way as to cause continuous charging, thereby keeping the cell at a 100% charge without causing cell damage.

d. The electrolyte is the ionic conductor solution that permits the transfer of electrons from the positive pole of a chemical cell to the negative pole.

e. The memory effect is most often found in those ni-cad cells which are used in standby service. This condition occurs when a cell is discharged a number of times to only a small percentage of its service capacity. Somehow, the cell "remembers" the level of its earlier discharges and will not provide a greater service capacity than the "remembered" level.

List of Illustrations